Burçe Çelik did her BA in Communications at Anadolu University. After working in private television, she started her academic career at Yeditepe University where she also did her MA in Communications. She received her doctorate degree from McGill University, Montreal. She has been writing on cellular telephony and the social history of communication technologies, such as cassette technologies and the telephone in Turkey, in both Turkish and English. She currently teaches in the Faculty of Communications at Bahcesehir University, Istanbul.

TECHNOLOGY AND NATIONAL IDENTITY IN TURKEY

Mobile Communications and the Evolution of a Post-Ottoman Nation

BURÇE ÇELİK

I.B. TAURIS
LONDON · NEW YORK

Published in 2011 by I.B.Tauris & Co. Ltd
6 Salem Road, London W2 4BU
175 Fifth Avenue, New York NY 10010
www.ibtauris.com

Distributed in the United States and Canada
Exclusively by Palgrave Macmillan
175 Fifth Avenue, New York NY 10010

International Library of Cultural Studies 15

ISBN 978 1 84885 429 1

A full CIP record for this book is available from the British Library
A full CIP record for this book is available from the Library of Congress

Library of Congress catalog card: available

Printed and bound by CPI Group (UK) Ltd, Croydon, CR0 4YY
Camera-ready copy-edited and supplied by the author

FSC
www.fsc.org
MIX
Paper from responsible sources
FSC® C013604

CONTENTS

PREFACE

Cellular telephone technology was adopted in the so-called developing countries more quickly and passionately than in developed countries. This was true in a developing Turkey. Cellular telephony in contemporary Turkey defines a social practice, collective attachment and even an addiction of sorts. This study produces a possibility of understanding why and how the cellular telephone has become so popular and an object of collective attachment/addiction in Turkey. This book aims to produce a social study of technology, in particular Turkish cellular telephony, by attending to the social, political, economic, cultural, historical and affective conditions within which cellular telephony as a social practice emerged in Turkey.

In this work, I attempt to understand why and how the cellular telephone has become a collectively attached and widely imitated technology in Turkey by exploring its performance in relation to the possible conditions giving rise to the emergence of cellular telephony as a social practice. I situate the performance of cellular telephony not only on the basis of its obvious meanings and visible instrumental functions but also via the desires, imaginations, inclinations, wishes, purposes, sensations that it responds to and reproduces. By drawing relevant arguments from the philosophy of technology, psychoanalysis, sociology, post-phenomenology, Turkish critical studies and literature and by benefiting from resources such as cellular telephone users, print, visual and electronic media, this work produces a possibility of understanding why and how cellular telephony becomes an appealing and an attachable technology for marginalized bodies.

ACKNOWLEDGEMENTS

First of all I would like to thank Dr. Jenny Burman who has believed in this project from the very beginning and has done more than supervising this work by giving encouraging comments on my work that motivated me to think and write with a great enthusiasm especially in the last intensive months. I am also thankful to all of my informants who spared their time for this research and have openly shared their thoughts, secrets and feelings with me. This study could not have been carried out without their collaboration. Mehmet F. Toprak and Devrim Biltan helped me to find the database search program for print media search and to reach the GSM networks' television advertisements where accessibility to old resources is very much limited.

I am indebted to my dear friend Defne Karaosmanoğlu for reviewing drafts of the manuscript, giving constructive criticisms and also for her enthusiastic and encouraging attitude to my work. I benefited a lot from advice and inspirational discussions on various issues relating to technologies, the imagined gaze of the west, melancholia, cellular telephony, the recent history of Turkey with Isaac Isitan, Bashar Shbib, Dr. Meltem Ahıska, Dr. Tül Akbal Süalp, Tuğçe Göktekin, Carole Poliquin, Elfin Yüksektepe, Dr. Kaya Özkaracalar, Aynur Celik, Baflak Doğa Temür and Tolga Kutluay. I would also like to express my thanks to Andrea Braithwaite for her careful and detailed editing of the text and also for her thought-provoking comments. Also I am indebted to Tanzer Ercanpolat who designed the final lay-out of the book.

Finally I would like to thank my son Kerem, my parents Nedime-Abdullah and my sister Beliz for their love, support, and patience.

1

INTRODUCTION: CELLULAR TELEPHONY, IMITATION, ATTACHMENT

In Turkey, as in many other countries around the world, cellular phone technology has grown incredibly popular and become an object of collective attachment. Cell phone technology was introduced in 1994, when Turkey was experiencing one of the most major economic crises in its history. Despite the economic depression, the cell phone market continues to expand. In the early 2000s, global cell phone manufacturers have declared Turkey as a jewel of the world market, a country that promises massive profits, as its huge, young population is eager to engage with technological novelties.[1] Turkish industry giants, initially hesitant to invest in cellular telephony in the late 1990s, attempted to carve out shares of this rapidly and unexpectedly growing market in the early 2000s, by announcing that they could not foresee the boom in cellular telephony yet were now ready to play a role.[2] By the early 2000s, possessing and using a cell phone had become a pre-condition for people to be 'there' in social space.

In 2010, cellular telephony is a large industrial and technological complex with three GSM networks, 67 million users[3] with diverse demographic characteristics, base stations that cover almost the entire country and divide it into communicable cells, and approximately 100 million machines[4] in active use (some of which are adaptable to the third generation mobile system). Importantly, cellular telephony also describes a collective practice of national attachment in contemporary

Turkey. Today, the cell phone is often described as Turkey's 'national organ'[5] in journalistic representations, integrating a society otherwise fragmented into enclaves of identity politics into a 'nation'.

Lewis Mumford (1963) presciently noted that understanding technology brings us to an understanding of society, of the people who make and use technology. In this book, I produce a persuasive understanding of why and how the cell phone has become a popular object of collective attachment in Turkey. By thinking of cellular telephony as the most favoured technological practice in contemporary Turkey, I herein reflect on the 'present' of Turkey, uncovering in the process the historical, affective, cultural, social, economic and material conditions that transformed cell phone from a tool into a technology – one to which many have become attached and even addicted. Contending that technologies consist of more than their material and mechanical work, that they are also bound up in desires, inclinations, affectivities, imaginations, responses and sensations in their performance, I ask what is wrapped in cellular telephony in Turkey? What kinds of collective desires and imaginations are expressed in its performance? What kind of alarm does it have that many people of Turkey cannot stand not paying attention to? I consider the performance of cellular telephony in Turkey as a response to a collective's conscious and unconscious call. By situating cellular telephony as a response, I can thus comment on the 'call' of the collective that has remade a tool into an attached and addicted performance.[6]

This book aims to produce a social study of technology, in particular Turkish cellular telephony, by attending to the social, political, economic, cultural, historical and affective conditions from within cellular telephony as a social practice has emerged in Turkey. Social studies of technology begin by questioning the social, economic, cultural, political and economic conditions that give rise to technological invention and change (see, e.g., Bijker et. al. 1987, Law and Bijker's canonical 1992 work). Since technological inventions are products of human beings and are designed to serve human purposes, the technological change is always conditioned by factors in the human environment. As Mumford has pointed out, technics are never self-contained, but always react to the impulses and forces from even remote parts of the environment (1963: 6). Thus if we are to question technological change, we must start by questioning the environment, the social, cultural, economic and political conditions that give rise to such change. Even though I am not addressing an invention of Turkey, the cell phone is domesticated and imitated in Turkey, and it has become a social practice with which many people are engaged. It has mutated from a foreign technology to a 'national organ'. This shift requires us to think through the conditions that make such appropriation possible – what are the material, social, cultural, historical, political and affective contexts that enable the domestication of the cell phone in Turkey?

Many studies have shown that the cell phone is popular worldwide, and that it was adopted more quickly and passionately by marginalized individuals, groups and countries (Schwartz, 1996; Agar, 2003; Katz and Aakhus, 2002). Not only has this technology moved throughout the world, but it has also spread more quickly to places whose relations to globality are characterized by economic, political, and cultural asymmetries. Recent ethnographic research has revealed how different collectives find their own use for the cell phone (Donner, 2005a; Horst and Miller, 2005). As a global technology, the cell phone has different meanings, tasks, performances, and uses and it produces different practices in different parts of the world. These empirical studies all indicate that cell phone technology is highly flexible, that it can be used for a variety of purposes which change across social contexts (Plant, 2000; Chesher, 2007). Sadie Plant (2000), whose study investigates cellular telephony on a global scale suggests that the cell phone is capable of satisfying diverse aims and can be used in a variety of cultural and social contexts. The cell phone traverses localities, and differs depending on where it is used, produced, invented or imitated. When we speak of the cell phone as an invention, we speak of a particular technological device that enables both point-to-point and mass communication between bodies that are mobile and distant from each other, but as a social practice we speak of multiplicities, varieties, and differences.

Gabriel Tarde's (1962) concept of imitation is useful for approaching the varieties that come out of a spread of an invention,[7] as well as for exploring the reasons behind an invention's imitation. Although in his early writings Tarde implies that imitation can only occur via face-to-face communication, his later work suggests that imitation happens without any place-based limitations. For Tarde, an invention always spreads through imitation, and imitation is what lends social character to an invention (1962: 382). As an invention is imitated, it multiplies, changes, and grows distinct. Tarde points to logical and extra-logical laws that determine whether or not an invention is imitated. Logical laws explain that an invention is imitated to create increased utility, or because the invention is thought to solve a problem better than others. Imitation is most often caused, however, by extra-logical incentives that are not rooted in rational or deliberate consideration. For Tarde, extra-logical incentives include cultural, emotive, and affective forces that make imitation promising and meaningful.

When we speak of cellular telephony becoming localized, and of social practice as a sort of imitation, then we need to ask about the causes that make cellular telephony imitable in certain contexts, particularly in marginalized spaces or by marginalized groups. The rapidly expanding literature on cellular telephony offers some reasons for its wide imitation in changing contexts by showing how, where, when, by whom and for what purpose the cell phone is used. The majority of this research is based upon audience reception theory and its methodologies, and so sees the cell phone as a device that provides higher utility and better problem solving than other contemporary technologies. Such explanations rarely take into account macro socio-technical contexts, focusing solely on observations about the material, temporal, and physical conditions that make the cell phone more instrumental, and on those for whom it is more useful.

Following Tarde's understanding of imitation, the majority of research on cellular telephony has thus uncovered the logical laws that make cellular telephony an imitable technological practice. The reason for its wide imitation has often been posited as a consequence of its instrumental value. Empirical studies whose findings are based on fieldwork demonstrate the cell phone's use and instrumental value for people who are physically mobile and need instantaneous and spontaneous connections with others. Richard Ling and Brigitte Yttri (2002), prominent cellular telephony scholars, state that cellular telephony is characterized by three types of activities: security, coordination and social interaction (see also Ling and Yttri, 1999; Katz, 2001; Cooper et al. 2002). Illustrating how cellular telephony is generally interpreted as an amalgamation of its visible functions, they argue that the cell phone is used primarily for coordination and social interaction, and point to its role in practical matters like organizing daily routines or changing plans, emphasizing its growing embeddedness in the patterns of daily life. The cell phone is also thought of as a medium, one which increases the user's sense of security because it enables connection almost anywhere, at any time (see Katz and Aakhus, 2002; Garcia-Montes et al, 2006; Ito, 2005). According to these studies, then, what makes a cell phone an imitable technology is its instrumentality to urban lifestyles in which mobility and risk are considered everyday and organizing principles.

Similarly, the popularity of cellular telephony in so-called developing countries has been investigated largely by looking at these regions' material circumstances. For instance, cellular telephony has become popular in areas where the telecommunications infrastructure does not work properly or reliably (Agar, 2003; Schwartz, 1996; Best, 2003), due in part to the fact that cellular telephony does not require a substantial investment in infrastructure. In places where the mediascape falls under strict state control, cellular telephony functions as an alternative mass media, enabling the circulation of messages among subscribers (see Lallana, 2004; Hill, 2003; Agar, 2003). As texting via cell phone is more affordable than most other means of connection through communication technologies, it has become popular in countries where poverty prevails (Pertierra, 2005; Elwood-Clayton, 2005;[8] Strom, 2002; Goggin, 2006).

All of these studies are valuable in the sense that they reveal how logical laws govern the imitation of cellular telephony in particular contexts. Reviewing Tarde's theory, Christian Borch (2005) proposes an analytical focus on the structures that condition and regulate imitation, while at the same time attending to the contingency and variability of imitation dynamics. He recommends an emphasis on how spatial designs are applied in order to govern imitation at a distance, through rhythms. Following Borch's directive, we can also include the lack of public telephones in public spaces, the chaotic order of automobile traffic in cityscapes which unpredictably extends waiting time, the lack of railway systems to ease the rush-hour traffic, and the cacophony of streets in cities to an analysis of the imitation of cellular telephony. These common scenarios contribute to the impression of cellular telephony as highly useful technology that speaks to the problems of everyday life better than others. As James Katz (1999) has written, people use cellular telephony in traffic to 'kill the waiting time' by doing something else and also as some other authors suggested the cell phone 'resurrects' mobile time that would have previously been considered economically unproductive or dead time (Perry et al. 2001; Green, 2002; Townsend, 2001). Ling and Yttri (2002) discuss how people use cellular telephony to ensure easy coordination: when meeting in a certain place at a certain time is always subject to change due to traffic, people are more likely to perceive a need or use for a cell phone in order to re-coordinate. Rural areas also typically lack public telephones in public spaces, and cellular telephony has proved useful for peasants or fishermen that are relatively far from markets in facilitating the flow of information between city and rural areas (Tall, 2004; Samuel et al. 2005). These sorts of reasons augment the popular perception of the cell phone as an instrumental and therefore imitable invention.

These are also elements that begin to explain why and how cellular telephony is imitated so widely and so easily in Turkey. I propose to be as thorough in my approach to Turkey's collective attachment to and imitation of cellular telephony, for as Tarde notes, imitation is most often rooted in non-rational or extra-logical laws. Questions of how the cell phone is imagined and fantasized, how it moves people, why one wants to make or receive a call, and what kinds of affectivities or emotionalities this contact produces remain under-theorized and under-analyzed in these studies. The literature on cellular telephony seems to be concerned with exploring the active uses of cell phones in everyday life and in different contexts; in return its emphasis tends towards cell phones' instrumental functions. Accordingly these studies document information, ethnographic data, and fieldwork findings about the technology's instrumental value, but do not furnish us with analytical tools through which we can understand how the cell phone becomes an object of collective addiction.

When casting about for causes that are not rooted in rational incentives, current literature on cellular telephony generally adopts a view that considers technologies as representations, meanings, or symbols. Positing that cellular telephony is invested with a social promise, the majority of this research offers only a limited analysis of Tarde's extra-logical laws. Rather, this work rests on Ervin Gofmann's view of the social, which can be simply defined as mutual susceptibility and so examines cell phones as a social medium that manages how individual bodies appear to others. Accordingly, cellular telephony's social promise becomes seen as providing a means of social acceptance, through implying social status and particular lifestyles to polish one's social face and gain recognition or popularity within existing social relations (see, e.g., Ling, 2003; Oksman and Rautuaninen, 2003). This perspective is shared by most of the scholars who discuss cellular telephony in Turkey; the bulk of the studies in this limited literature suggest that cell phones in Turkey 'show' one's social status, or a desired status which is not currently held (Özcan and Koçak, 2003; Geray, 2001). The cell phone is treated as a commodity, whose purchase brings a sense of upward social mobility to its users. For instance, cell phones' early connotation of being a 'rich men's toy' (Katz and Sugiyama, 2005), is argued as the source of its attraction for the 'poor' and 'minorities' (Katz and Aakhus, 2002; Sheller, 2004: 42) who are invested in the notion of social mobility.

The cell phone as an object thus has a display value, which is assumed to reflect its holder's social status. Studies have also proposed that 'developing countries' associate the cell phone with western and modern values. Pajnik and Lesjak-Tusek (2002) explore how images of 'modernity' and 'western values' are referenced interchangeably in Slovenia's ads for cell phones. Zhang and Harwood (2004) show that in China ads for communication technologies draw heavily on the notion of 'modern' (see also Vershinkaya, 2002 (Russia); Chakraborty, 2004 (Bangladesh)). All of these studies have considered the cell phone as an object that gives meaning on behalf of and for its user, as a presentation of one's self, identity and/or social face (Katz, 2002; Lobet-Maris, 2003; Ling and Yttri, 1999; Fortunati, 2001; Donner, 2005a).

According to these studies, the cell phone is imitated in particular contexts because it provides higher utility for people who needed a more mobile means of communication, or because it functions as a sign that reflects the user's desired social status. The imaginative, psychic, affective or sensory aspects of cell phone technology are dismissed or ignored altogether, in accordance with audience-reception theory and methods. The terms of this theory limits cellular telephony to a technological gadget that 'represents' certain values and fulfils certain functions for its users. Cell phones seem to have no other role in people's lives beyond these meanings and instrumental functions, no intrinsic relation to the human body outside of its represented values, no anchor in fantasies, imaginations, responses, or conscious and unconscious inclinations. Audience reception theory only allows for an exploration of the actual and/or practical uses of cellular telephony; researchers have no framework for linking the concrete roles to the macro socio-technical context in which cellular telephony is widely imitated and becomes a social practice.

In this work, I attempt to understand why and how the cell phone has become a collectively attached and widely imitated technology in Turkey, by exploring its performance in relation to the conditions of possibility that give rise to the emergence of cellular telephony as a social practice. I situate the performance of cellular telephony not only on the basis of its obvious meanings and visible instrumental functions but also via the desires, imaginations, inclinations, wishes, and purposes, sensations that it responds to and reproduces. Avital Ronell (1989) writes that the telephone is both lesser and greater than itself. It is lesser in the sense that it is a machine that enables point-to-point communication. But it is also greater than itself as its performances exceed its material work in the imaginary. By paralleling the material performance of the telephone to the maternal performance of the mother, Ronell reveals the invisible links between the body and the telephone, as a synecdoche of technology. These invisible links are the sensations, affectivities, imaginations, hallucinations, and illusions that push the telephone's function beyond the bounds of its technical assemblage. In this work, I argue that the cell phone, as a much more fluid and flexible technology than the telephone, performs in ways that transcend its instrumental and symbolic value. The cell phone impresses, affects, moves and gathers bodies in particular ways that incur imitation and collective attachment.

When searching for the reasons behind Turkey's imitation of cellular telephony, I deliberately pay sustained attention to the 'extra-logical' laws that cause this invention to be imitated. This approach makes it possible for me to provide compelling accounts of why and how the cell phone has become an object of collective attachment or addiction. Attachment and addiction define processes through which cellular telephony transforms from a technology that is used and imitated to an object whose lack makes bodies feel depressed, anxious, or lonely. Thus understanding the performance of cellular telephony in Turkey is crucial for understanding why and how it has become a collective attachment for millions of people. Yet a technology takes on more than just visible or functional tasks (see, for instance, Ronell 1989; Verbeek 2005). The bond that arises between people and artefacts is based on how the concrete object affects and produce sensations, and 'not only the meanings or symbols it carries or the functions it fulfils' (Verbeek, 2005: 225). A technological artefact colonizes a portion of its users' inner lives, always serving and servicing desires, dreams, unconscious and conscious purposes (Sofia, 2000). It becomes a user-friendly product insofar as it responds to these wishes (ibid.). It is in these ways that imitated technology is instrumental and meaningful, and not solely because its functional aspects seem to ease life practices. In considering the performance of the cell phone as something that produces and becomes a response to affectivities, imaginations, sensations, and emotions, my most pressing question becomes, how does it perform, how does it affect, move and gather bodies together in Turkey?

I understand cellular telephony first and foremost as a bodily practice that becomes social through its imitation by individual bodies. In his theory of techniques of the body, Marcel Mauss (1973) tells us that the body, the 'total man' and his actions need to be understood through a triple rather than single consideration. This tripartite approach requires seeing the body as a physiological, psychological and social instrument. Thus to regard cellular telephony as a bodily practice, we need to incorporate the cell phone's social, psychological and material promises alongside the conditions of possibility for cellular telephony to become a social practice. By drawing relevant ideas from the philosophy of technology, sociology of technology, sociology, psychoanalysis, poststructuralism, post-phenomenology, and Turkish studies I produce a framework for exploring cellular telephony in Turkey. In doing so, I have been influenced by some exceptional studies on cellular telephony, which emphasize the role of sensing, affecting and affected, and fantasies of the body and the collective in contemplating the bond between body and cell phone (see Richardson, 2007; Rafael, 2003; de Souza e Silva, 2005). In particular, Vicente Rafael's work on cellular telephony in the Philippines has proved inspirational. His discussion of the cell phone in relation to the historical and collective imaginations of technologies, and his interpretation of the cell phone mania in a social structure and space that is fragmented into enclaves or collectives on the bases of language differences, ethnic and class disparities, has been influential in my own analysis of cell phones' social and political promises.

The performance of cellular telephony cannot be dissociated from the conditions that make the cell phone an attachable technological performance. Many studies speak of the material and physical conditions that make the cell phone a highly useful and instrumental technology. Others have noted how socio-political contexts import significance to the technology, surmising that it might foster political and social movements that could transform social realities. However, these need to be considered in tandem with the affective and historical conditions that enter into the ways in which the cell phone performs, as a means by which this performance becomes appealing, imitable and attachable. The same technological object, with the same use value, might have different meanings and promise different pleasures according to the affective contexts they inhabit (see Grossberg, 1990). Likewise, an analysis of how technologies have historically been perceived, interpreted, used and imagined in Turkey and even in the Ottoman Empire is crucial to grasp why and how cellular telephony has become popular and collectively attached in Turkey. I argue that melancholia, both as a feeling that dominates everyday life and as a psychic and social condition of the bodies that dwell in Turkey, determines the way people engage with technologies. I also rethink the concept of technoscape, to develop an approach that elicits the imaginational and emotional experiences of technologies, and addresses how history and social structure enter into bodies to influence their movements and actions in social space. Incorporating insights from Turkish literature, critical studies and social studies, I identify and comment on the historical, social, cultural, affective, political and economic conditions that give rise to cellular telephony in Turkey.

I use a variety of resources to provide a detailed and nuanced understanding of cellular telephony in Turkey. My research is largely empirical in character, and includes print media research, analyses of advertisements for GSM networks and in-depth interviews with cell phone users (and also with people who produce GSM networks' ads, scriptwriters, commercial directors who in one way or another have influences on the ways in which cellular telephony is imagined in

Turkey). All of these sources, and especially the interviews with cell phone users, provide valuable hints about how cellular telephony is imagined and what is bound up in its performance in terms of collective desires, responses, purposes, wishes, and affectivities. However, this material on its own does not fully reveal the intricacy of cellular telephony. As such, my study also contains discursive and interpretive strategies, which in some ways make my discussion of cell phone performance speculative while grounded in established psychoanalytic and poststructuralist theories. My approach to cellular telephony extends past examining and recording how, where, by whom, and with what conscious purpose the technology is used, to situate and understand its work in the social imaginary; speculation is thus inevitable and necessary in order to see the machine's transformation into a collectively attached performance. To this end, my resources have offered substantial insights for producing a possible and plausible understanding of cellular telephony in Turkey.

As part of my media research, I examined all national print media, including newspapers, magazines and journals, between 1997 and 2007. I used a local database search program which allowed me to scan all the news, articles, cartoons, and essays about cellular telephony published over the past decade. I also examined three national newspapers, namely *Hürriyet*, *Milliyet* and *Zaman*, for the period between 1994–97, searching for news, articles, and essays that one way or another touch on the issue of cellular telephony in Turkey. In addition to print media, I explored three popular websites for entries that relate to cellular telephony in Turkey. These three popular sites, namely ekşisözlük (a sort of youth community website), agora (blogger site of *Hürriyet*,) and itiraf.com (confession board), have the potential to reflect how cellular telephony is imagined and experienced in Turkey, especially by youth. Each of these websites has detailed sections on cellular telephony where bloggers recount stories, express opinions, and discuss cellular telephony in Turkey. I also collected and examined GSM networks' television and print advertisements from 2000 onwards. Television viewing is a significant social practice in Turkey, and many people regardless of their demographic differences watch local television. Television ads in particular reveal the various imaginations and desires imposed on and invested in cellular telephony.

My interviewees are the most significant resource in my research. I conducted 42 interviews with cell phone users covering a wide range of demographic characteristics. I used the snowball method to find participants, yet I paid close attention to ensure a sample that reflects Turkey's diversity in terms of ethnicity, political orientation, class differences, age, gender, and occupation. All my interviews took place in Ankara and Istanbul, and I also interviewed people who have recently moved to Ankara from smaller villages. Cell phone use is especially popular among teenagers in Turkey, as it is in almost every region of the world, and so teenagers over the age of 14 have also been included in my sample. I conducted additional interviews and focus group discussions with interviewees who agreed to talk at length with different people. In these focus group conversations I tried to expand the discussion of cellular telephony, and moderated to encourage the group towards conversations among themselves about their experiences with cellular telephony.

The following two chapters explore the historical, affective, cultural, social, political and economic conditions that I argue instigated the collective attachment to cellular telephony in Turkey. Chapter two offers a nuanced interpretation of Turkey's history, aligning key historical breaking points with a discussion of melancholia. Critical studies in Turkey have investigated the ways in which a Turkish national identity has been imagined by policy makers and state bureaucrats throughout the country's modernization period. Yet, the question of how these imaginations have effected and affected the way people live, experience and imagine modern Turkey has received little scholarly attention. In an effort to explore Turkey's historical and affective context, this chapter establishes melancholia as a primary sentiment in everyday life, as well as the psychic and social condition of the individual and the collective body. Exploring melancholia in its historical context will provide an understanding of how technology – in the broader sense of technique and of scientific and positivist thought – has been imagined, experienced and acquired meaning in this landscape.

Chapter three deals specifically with the notion of the technoscape, and develops a theoretical approach to understand cellular telephony as a contemporary performance of the technoscape. After providing historical and introductory accounts of key dynamics and issues with regards to technologies in Turkey, I conclude by demonstrating why the technoscape in Turkey should be conceptualized as an historical dream of presence, outside of which cellular telephony cannot be effectively and affectively thought. This chapter explores how social structures and historical imaginations, emotions, desires, and wishes are central to the shape of technological practices. Drawing on arguments from contemporary discussions of the technoscape and globality, social studies of technology, and poststructuralist approaches to technologies that expose the logic of technological plays, this chapter forms the theoretical perspective of my thesis.

From chapter four onwards, I interpret the performance of cellular telephony in Turkey. Chapters four, five and six are particularly concerned with how cellular telephony is imagined and experienced in Turkey. In these sections, I elucidate and comment on the performance of the cell phone in Turkey in the light of my fieldwork and media research. In chapter four, I discuss the nature of the collective attachment to cellular telephony. Following a brief review of the current literature on cellular telephony, I show that the collective attachment to cellular telephony cannot be understood solely on the basis of its instrumental and symbolic value. Rather, I extend this discussion by debating whether the contemporary addiction to cellular telephony can be perceived as an addiction to consumption. After questioning, via Albert Borgman's analysis of devices, the utility of claiming that the cell phone is a commodity, I argue that the cell phone is a device that promises both the pleasures of consumption and the pleasures of producing a performance. In this chapter I also highlight the significance of the bodily experience of cell phones, as it comes to produce sensations and affectivities that exceed its work as a device.

Chapter five is concerned with the ways in which cellular telephony becomes a response to Turkey's collective desires, imaginations, inclinations, wishes and purposes. I draw on arguments from Zoe Sofia's analysis of containing technologies and Avital Ronell's analysis of telephony, emphasizing the logic of *fort/da* play to argue that the performance of cellular telephony in Turkey is most productively approached through the concepts of transference/translation and containment. Dealing specifically with issues like the indivi duation and articulation of singularities, this chapter suggests that the cell phone is a toy-like vehicle for Turkish bodies that feel out of place, as it provides a means of movement, exteriority, exit and an illusionary arrival at self-presence, home, or self-proper. In this chapter, I tease out the relation between melancholia as a psychic condition and cellular telephony as a contemporary dream of presence. Psychoanalysis provides a framework for grasping how the collective's bodily attachment to the cell phone is rooted in the psyche, as melancholia forecloses the means of individual articulation and establishes a sort of distance from the invading other(s); it creates conditions for fragmenting and polarizing the social sphere into political and social camps by sticking social categories into bodies. I maintain a critical view of psychoanalysis by employing historical, social and political contingencies in my analysis, as these factors are generally elided by the field's emphasis on the formative structures of familial and/or primordial relations.

Chapter six explores how cellular telephony, as a production of the social, is imagined and experienced in contemporary Turkey. Incorporating Elias Canetti's theories of the crowd, I propose that cellular telephony is a site in which collective desires for and imaginations of living in an open crowd are invested. Canetti writes that the reason for a crowd to emerge is the collective desire for sublime equality, justice. I discuss whether the cell phone – which is deeply embedded in social struggle – can be interpreted as a social technology which promises to generate an open crowd. By shifting the discussion of sociality into one of spatiality, I show that the dream of living in the crowd does not necessarily produce a site where all bodies can feel a sense of sublime equality and social connectivity, that the crowd is still a space where distances between bodies produce and preserve hierarchies, status and ranks. This chapter explores how cellular telephony as a social medium opens onto a hybrid space where networks and fluid spaces negotiate and circulate. By following Sheller's review of Harrison White's theory of social gel or goo, I discuss the ways in which bodies couple and decouple with each other through the cell phone.

This study offers one possible understanding of why and how cellular telephony has become an object of collective attachment in Turkey – a 'developing' country, one of the marginalized regions of globality, a space where the east and west are in constant negotiation. The framework I create throughout can also be used as a starting point for thinking about the unimagined popularity of cell phones in other marginalized countries or regions of the world. Rather than relying solely on social theory that sees cellular telephony in terms of its visible social effects and accomplishments, I attempt to grasp how it invisibly, unconsciously, and affectively responds to or even anticipates our collective wishes, desires, inclinations, and imaginations.

2

TECHNOLOGY OF/IN MAKING A MODERN NATION: MELANCHOLIC CONSTRUCT, MELANCHOLIC BODIES

Down to our day it has been a proven fact in human history that the members of states or other units of social organization that have lost their position of superiority – or their claims thereto – during the final days of their wars have usually had difficulty in reconciling themselves with the new situation, and most importantly, overcoming the sense of loss to their own self-esteem. A lengthy period – sometimes an entire century must often pass before this can be accomplished. And perhaps they will never succeed (Elias, 1972, cited in Akcam, 2004). Orhan Pamuk (2003) writes that the founding of the new country, the 'Turkish Republic', often seemed to him as showing a desire to get rid of the saddening memory of the objects that are mindful of the empire being lost, rather than of a joyful desire for modernization. Rather than mourning for the loss of the empire, this new construct called 'Turkey' simply rejected the empire (and rejected its loss) but could not escape from its remains, which is manifested in everyday life in the 'feeling of defeat and loss' and 'in pain of poverty'[1]. He often describes the feeling of this landscape as melancholy.[2, 3]

Similarly, Taner Akçam, a renowned historian, argues that the history of Turkey, especially around the turn of the twentieth century, has been a history of traumas. He catalogues massacres, genocide[4] and expulsions, as well as military defeats, territorial losses, and national humiliations at the hands of foreign powers. These traumas 'were so profound as to cause Turkish society to try and forget them, but they have left their scars on the national psyche just the same' (2004: 41).

In this chapter I propose to consider 'Turkey' as a melancholic construct from its foundation, and accordingly argue for an analytical sensitivity towards the historical melancholia that is part of the 'national identity' or the outcome of the 'interpellation' to Turkishness. In the following pages I discuss melancholia not only as a feeling of everyday life in Turkey, but also as a psychic and social condition of both the individual body and the collective. Exploring melancholia in its historical context will provide an understanding of how technology – in the broader sense of technique and of scientific and positivist thought – has been imagined, experienced and thereby acquired meaning in this landscape. The following discussion focuses specifically on two issues: how the Turkish Republic has historically imagined, idealized and located technologies within its melancholic imagination; and how cultural forms of melancholia contribute to foreclosures of sociality and individuality, as well as to commodification. These two trajectories will establish the analytical and historical ground needed for our analysis of contemporary cellular telephony in Turkey, by providing the historicity of melancholic imaginations of technologies and offering a nuanced view of the contemporary social, affective, and political conditions within which cellular telephony has become a collective attachment.

Introductory accounts of melancholia

According to Freud, after the loss of a loved object, the melancholic loses the capacity to love, her/his self-esteem, and all interest in the outside world. The melancholic simply withdraws to her/his inner world where s/he constructs a relation with the lost object. In contrast to the normal process of mourning after the loss of a loved one, melancholia consumes the melancholic's ego, leaving the subject with a heightened conscience or critical agency which is in constant battle with the ego. In melancholia, the libido for the lost object is not transferred to another object but regressed to the ego, and this regression generates the ego's narcissistic identification with the lost object. Thus the melancholic incorporates the lost object into her/his psychic domain, maintaining a connection to the loss in her/his unconscious.

In his famous essay, 'Mourning and Melancholia', Freud (2001) writes 'the patient is aware of the loss which has given rise to his melancholia, but only in the sense that he knows whom he has lost but not what he has lost in him' (2001: 245). He suggests that it is not possible to understand what absorbs the melancholic so painfully, because it is not possible to see clearly what has been lost. The loss in melancholia refers to a loss of more ideal kind that the melancholic cannot bear, s/he therefore pushes it into the unconscious to live there as part of the ego. The loss that gives rise to melancholia has been defined by Freud to include a social or political ideal – such as liberty or country – to which the body has a libidinal attachment. In order for an ideal to be lost, as Freud's writings demonstrate, it does not need to have been previously possessed. One can lose an ideal and thereby suffer because of its loss when social and political conditions do not allow for its presence as an object of attachment. This loss of more ideal kind, whether it refers to a loss without a lost object or a loss of what has never been possessed, gives rise to melancholia as the loss is incorporated or swallowed by the ego. Melancholia is, therefore, a psychic strategy of a body that cannot let go of the lost object from external reality but only transfer it into the internal world (Agamben, 1993). In this melancholic strategy, the loss is both affirmed and denied; melancholia essentially refers to an epistemic split (see Comay, 2006). As a strategy, melancholia succeeds in appropriating its own object only to the extent that it affirms its loss. Affirmation and denial work concomitantly to defend the melancholic from the trauma of recognizing the lost object's inaccessibility. Thus, affirmation of the loss is precisely for its appropriation within the ego, outside of which any relation with the lost object would be impossible. There is a recuperative logic of gain-through-loss in the project of melancholia (Comay, 2006: 89).[5]

The epistemic split created by simultaneously affirming and denying loss is assimilated into the melancholic's ambivalence towards the lost object. The object of melancholia provokes a strange battle in the unconscious, wherein 'countless separate struggles are carried over on the object in which love and hate contend with each other; the one seeks to detach the libido from the object, the other to maintain this position of the libido against the assault' (Freud, 2001: 256). The loss of an object precipitates this ambivalence, as part of the process of letting go (Butler, 1997: 175). Freud (2001) suggests that this psychic articulation of ambivalence is shaped as a conflict between one part of the ego and the critical agency. Freud makes clear that 'the accusations that the critical agency is said to level against the ego turn out to be very much like the accusations that the ego would have levelled against the object or the ideal that was lost' (2001: 248). He writes, 'we perceive that the self-reproaches are reproaches against a loved object which have been shifted away from it on to the patient's own ego' (ibid.). The critical agency judges the ego through taking the 'ego-ideal' as the measure for its judgment. In his later writings Freud identifies how this ideal, which binds narcissistic libido, has a specific social aspect in the sense that it is also a common ideal of a family, a class or a nation. In *Civilization and Its Discontents* (2002), Freud argues that social ideals are transformed into a feeling of guilt through internalization; guilt is engendered in melancholia due to one's aggression toward the ideal (2001). The impossibility of fulfilling the ideal is turned inwards and produces self-aggression, and this in return becomes the essential structure of conscience. Through identification with the ideal, the body internalizes the 'unattackable authority' into itself, into the conscience (Freud, 2001).

In *The Psychic Life of Power* (1997) Judith Butler highlights that melancholia is also an account of social not only psychic domains:

> [T]he account of melancholy is an account of how psychic and social domains are produced in relation to one another. As such, melancholy offers potential insight into how the boundaries of the social are instituted and maintained, not only at the expense of psychic life, but through binding psychic life into forms of melancholic ambivalence. (1997: 168)

Butler's theory encourages us to think of melancholia as a cultural and social implementation of power, taking place within the processes of bodily subjectification. According to her, loss emerges from forms of subjectification: when the body seeks a sign of its existence outside itself, it enters the realm of social categories, terms, and names that are not its own, and the subject comes into existence through subordination to these categories. For Butler, participating in the domain of social categories means incorporating social and political ideals/texts into psychic space. As an outcome of the process of internalizing social and political ideals, the body becomes 'wounded', its potential to become a viable subject of power ripped away. Following Althusser,[6] Butler calls this process an interpellation of the subject by power, by which the subject incorporates the ideals of power as ego-ideals of its own: 'the process of forming the subject is a process of rendering the terrorizing power of the state invisible – and effective – as the ideality of conscience' (191). The melancholic comes to be blinded to his/her own loss as if it were a natural, biological outcome of one's presence in the world, a consequence of belonging to a certain gender category, ethnicity or nation. If we are to understand subjectivity as a product of subjectification through discourse (telling and showing what the ideals, norms are), we need to take into consideration both the work and historical contingency of power. Butler repeatedly emphasizes that the subject is a social product, emerging out of its performances in relation to positions constructed by power.[7] Here, in full agreement with Butler's theory, I will argue that the melancholia of bodies in Turkey is the very product of the historical organization of hegemonic power. As this introduction to the concept of melancholia demonstrates, loss is inextricably linked to the work of power and its given ideals; with this basis established we can now continue discussing my aforementioned argument about the melancholic character of ideal Turkishness.

Melancholia and ideals for national identity in Turkey

For at least three decades, political science theorists, sociologists, cultural studies scholars and historians have been preoccupied with the issue of Turkish modernity and formation of Turkish national identity in and out of Turkish academia. This now-expansive literature largely accepts that the foundation of a new country, within the confines of nationalism, modernity and secularism, has forcefully demanded its citizens radically break with their Ottoman past. From the alphabet to the name of the country, from clothing to national music, almost all domains of public life were transformed with the foundation of the new country. These changes were accomplished with the purpose of instituting a new country based on tenets of nationalism and modernity. We shall begin discussing what nationalism meant for the early Republicanists in Turkey and how their policies worked to nationalize the country.

Many authors suggest that the Ottoman intellectuals and then early Republicanists had viewed the lack of nationalism in society as the reason for the collapse or disintegration of the empire. The new country, founded in 1923 to replace the Ottoman Empire, was therefore imagined on the basis of nationalism. As M. Kemal Atatürk, the leader of this movement and the first president of the new republic, said: 'We are a nation that has come very late to and acted very negligently in the implementation of nationalist ideas…Our nation has paid an especially bitter price for having arrived without knowing nationalism. We have (now) understood our error was that we have forgotten ourselves' (cited in Akçam, 2004: 63).

Instituting a country based on nationalism was the ideal, yet the Ottoman society did not solely consist of Turks or Muslims, it has always contained cultural, ethnic and religious diversity. Starting in the nineteenth century, the Ottoman's multinational landscape was forcefully transformed into a Turkic-Muslim population. Consequently, this nation state, comprised of 'Turks', was called 'Turkey' instead of the Ottoman Empire. According to historian Cemal Kafadar (1995), 'Turchia' was not a term by which Ottoman subjects identified themselves; the self-appellation 'Turk' derived from their relationship with Europe (cited in Navaro-Yashin, 2002: 10). Europeans referred to Ottomans as Turks, which often sounded humiliating to the Ottoman rulers and elites (Akcam, 2004). By rejecting anything Ottoman and replacing it with the new country's ideals – which were largely inspired by Europe's modern-nation-states – and with a new name, early Republicanists also attempted to Turkify their history. A nation-state needs to claim historical continuation in order to be legitimate in a world of nationalist discourse. Thus, the history of 'Turkey' was rewritten as the ongoing story of Turkic groups in ancient central Asia, excluding other ethnicities and religions (Navaro-Yashin, 2002; Akcam, 2004). By changing the alphabet from Ottoman script to Latin, access to the Ottoman history without specific qualification or education has become impossible.

On the other hand, authors suggest, the early Republicanists believed that the Ottoman Empire had disintegrated and finally collapsed because it could not adopt the means of civilization, modernity and rational thought. Ottoman intellectuals and then early Republicanists 'saw' the need to adapt to the requirements of the contemporary world, to embrace modernity, scientific-rational thought and positivism in order to be seen as a progressive country in the international arena. As ¢erif Mardin (1994) writes, Ottoman intellectuals of the 1860s imagined a society which would be liberated, one that 'did not yet exist in the Ottoman Empire', and this sort of utopia 'had become their (these intellectuals') natural somewhat uncomfortable habitat' (202). Such thinkers saw 'positivism' as the most crucial intellectual tradition the Ottomans lacked, a key factor in their declining international influence as positivist ideology had prevailed and brought prosperity to other European powers. Mardin continues, 'Turkish intellectuals of the 1920s, on the other hand, had to face a further problem that of relating positioning as utopians to the real citizens of the Turkey of the 1920s, namely to Anatolian peasants...They also had to relate either to the roots of the 'real' culture of Turkey, i.e. to Islam, or to a new imagined national identity, which was in the process of replacing Ottoman imperial identity' (203). The question was how to implement modernity, which was associated with positivism, rational thought and secularism, in a society whose life practices were largely organized in accordance with religious values and whose 'internal cohesion' conformed to canonic Islamic values (1994: 209).

The history of modernization in Turkey is ambiguous and complex, and cannot be reduced to a homologous, uniform understanding of modernity (Kasaba and Erdoğan, 2000). The ways in which state organizations envisioned and attempted modernization, and how modernity was imagined by rulers and elites, have always varied in relation to questions of the means by which Turkey would be modernized, and on what basis its modernity would claim continuity with history. However, existing literature tends to homogenize the diverse processes of modernization and related policies of the state. Homogenizing works largely partake in an argument which blindly accepts the orientalist view that Turkish modernity has taken western modernity as a 'model' for itself, and in return generated an insufficient 'copy' (Lewis, 1969). Some critics argue that modernization from above has simply repressed the 'real' culture and replaced it with an 'official' one, which in one way or another mimicked European models (see Keyder, 1997; Kadıoğlu, 1996). For instance, Keyder writes, 'modernization from above came to mean modernization of the solitary nation but not of its individual members' (1997: 45-46). From this perspective, the individuals' 'real' identity was stifled and always remained foreign to or outside of the imagined formation of national identity under the strict policies of westernization/modernization. However, within the framework of melancholia I have outlined, it is not possible to agree with this reasoning, as an identity cannot emerge outside of its relation with power. Thus, one way or another, power is implemented in the subject in order for the subject to become.

Indeed, some insightful analyses strictly oppose a reductionist understanding of modernization processes in Turkey which equates modernization with westernization and formations of national identity with an opposition between 'real' and 'official' (Kasaba and Erdoğan, 2002; Navaro-Yashin, 2003; Ahıska, 2005; Göle, 1990). Navaro-Yashin, for instance, argues that no 'real' or 'original' identity or culture can be found outside westernization or relations with the west. By underlining the cultural exchange between the Ottoman and Europe throughout history, she argues that these two entities have always intermediated each other at all levels; there is no possibility of Ottoman history/culture outside of its relation with Europe, just as European history/culture always includes its relation with the Ottoman. Similarly, Meltem Ahıska proposes we consider the formulation of imagined Turkish national identity by taking into account the ambiguities of and conflicts with what she terms the imagined western gaze. Like Navaro-Yashin, she suggests that the west has always been integral to the formation of national identity, that it is not possible to think of imagined national identity without acknowledging its intrinsic relation to the imagined western gaze. Not only was the west taken as a model for the formation of imagined national identity but it was also perceived as a threat to 'indigenous' values (see below). As Ahıska's analysis shows, however, this paradoxical projection of the west both as a model and a threat cannot be generalized as a consistent standpoint of the rulers, intellectuals and bureaucrats who, across at least two centuries, worked at modernizing the Ottoman and Turkish landscape. For instance, Kandiyoti (1997) shows us that while in the nineteenth century, intellectuals held up the 'culture' of Anadolu[8] as an authentic culture that could be reconciled with the means of ideal modernity, in the twentieth century, Anatolian culture started to be seen as backward and irreconcilable with modernity. Clearly the ideology and policies of modernizing military state power cannot be understood as consistent and homogenous. An analysis of modernity, national identity and the military state power needs to be sensitive to ambiguities, uncertainties, complexities and contingent relations. The following analysis takes the early period of the republic into consideration, as I identify and explore the ways in which national identity was imagined, including both its ambiguities and how the people of Turkey participated in this imagination.

The transition from the Ottoman Empire to the Turkish Republic has been a painful process. It has not only meant the loss of the empire as a symbol of power but also the erosion of various elements of Ottoman public life (such as the non-Muslim communities). Although discourses of 'ideality' (what the ideal for this nation would be) were ambiguous and contradictory – because there was no revolutionary theory behind the reforms and imaginations of ideality (Mardin, 1994) – one thing was certain: a break from the Ottoman past was necessary. The new imagined national identity can be roughly defined as 'Muslim-born yet secular, different from Arabs and keen to adopt western life practices and technologies' (Navaro-Yashin, 2002: 48), while faithful to indigenous cultures and traditions as the base of national culture (Ahıska, 2005). As this definition illustrates, the national ideal was conflicted – identifying oneself as Muslim traditionally meant to live according to the rules of Sharia, so how would it be possible to reconcile Islam and secularism? Remaining faithful to indigenous roots (which have always varied, Anadolu or not) while adopting western life practices also seemed antithetical, as western life practices were considered threatening to the pre-existing national culture (individualism and consumerism in particular were seen as endangering familial relations and community spirit). Yet a radical break with the Ottoman Empire was necessary, not just because the Ottoman was not a nation state or modern construct, but also to establish distance from a past that sustained feelings of backwardness, powerlessness and humiliation in the international arena (Akçam, 2004). Or as Leyla Neyzi wrote, the ideals were to be 'achieved by cutting ties with the Ottoman past, the world of elders' (2001: 416).

Accordingly, under this new – young – construct, any reminders of the Ottoman past were rejected or negated. While not everything Ottoman was prohibited (though, for instance, religious schools were closed and wearing the fez forbidden), most Ottoman cultural practices were disavowed by being labelled as backward, or deliberately forgotten as a 'foreign' history. In doing so, the possibility of recognizing loss was also destroyed. Instead, railroads were designed, schools were built, the postal system was changed, the schooling system was transformed and borders were patrolled (Navaro-Yashin, 2002: 12). All these transformations were seen as fundamental for achieving the dream of modernity (see also Davison, 1990; Demir, 2005). The crucial question, however, is how did the people of Turkey interpret all these transformations, changes and expectations of the elites or ruling class and how did interpellation to Turkishness engender becoming life practices for their part? Although we have extensive literature detailing macro perspectives on how modernity was imagined and how nationalism was formulated by the state and its bureaucrats, we have very few analyses of the ways in which citizens reacted, received and participated in the formation of new country in the early period (see also Mardin's critiques, 1997). ¢erif Mardin's (1990) and Meltem Ahıska's works might be helpful for us to have an idea about people's lives at the time (rather than how the state formulated modernism), as Mardin looks at 1940s Turkey, and Ahıska covers the 1930s to late 1940s.

Both authors draw our attention to the institution of the uncomfortable habitat – which will become clearer below – on the part of citizens of this new country, made visible in the ways they acted, moved, behaved and related with each other and with the new ideals of country. For Mardin, 1940s Turkey was characterized by a climate of 'fairy-tale' in which people were implicitly but effectively expected to act 'as if' the ideals were fully adopted, 'as if' the modernization and nationalization policies generated meaningful practices. In other words, acting 'as if' one believes in what one was doing became the 'habitus'[9] of people of the 1940s. What is striking in Mardin's analysis is how discourses (in the strict sense as narrations) were produced and circulated in such a way that nobody really believed in what they were saying or hearing, yet people carried on 'as if' they believed in what they were doing. For example, as workbooks were written in

Latin alphabets, neither students nor teachers were able to read or understand what was written, and still the entire education system continued to work 'as if' the material was accessible, 'as if' there was actual schooling going on. Even bureaucrats working for the education ministry acted as if the system was working properly, although they knew very well that this was not the case. In short even though nobody believed in what they were doing, they continued to do it 'as if' the work was meaningful.

Education has played a significant role in the Turkish social engineering project, engendering a nationalist consciousness and a desire for modernity in children and youth.[10] Before s/he was a 'self' or part of her/his family, a child was the object of the state's nationalist and modernist pedagogy. The 'oath' recited – still today – by Turkish schoolchildren every morning goes as follows: 'I am a Turk, upright, diligent. My law is to respect my elders, protect those younger than myself. To love my country and nation more than my own self. My ideal is rise up and go forward. Let my being be sacrificed for the sake of Turkish existence' (cited in Neyzi, 2001: 416). As Leyzi insightfully writes, the 'oath' clearly shows that 'the "self" of the republic becomes the "self" of the individual, who must be prepared to sacrifice himself or herself for the nation' (2001: 417).

Ahıska's (2005) study of Turkey's early radio broadcasting delineates how radio worked as discursive practice for state power in its effort to modernize the country. In her work, radio appears as a technology of fantasies, offering a utopia of nationalized-modernity as if that had been the social reality of Turkey in the 1930s and 1940s. Ahıska argues that the performance of radio broadcasting worked to create the image 'as if' 'modernity ideals' had been achieved and 'as if' the nation had fully adopted the means of a nationalized-modernity. After conducting multiple interviews with early radio personnel (program producers, technicians, directors, writers), she suggests that the people who worked in radio at the time were far from believing in what they depicted as the 'reality' of Turkey. Her analyses highlight how these employees knew that 'nationalized-modernity' was clearly a fantasy, yet continued to act 'as if' it was the reality of Turkish public life. Ahıska shows us that neither the formation of the fantasy nor the persistence of military power of the state were separable from the public life of the 1940s. Strict state policies of promoting fantasies as real, Ahıska suggests, bifurcated reality into two domains. In the domain of external reality, people acted 'as if' they had lost nothing and the fantasy was real, 'as if' these discourses of modernization were meaningful. On the other hand, in intimate spheres, what Ahıska terms inner reality, where people feel more comfortable expressing their feelings and private worlds, the lack of belief and the sense of a 'wound' became decipherable. However, rather than understanding these two domains as 'official' versus 'real culture', Ahıska suggests, both worked concomitantly to perpetuate the fantasy. The 'excess' of imagined national identity was being worked through in this way (by perpetuating two domains of reality) to sustain the hegemonic power of the state. She calls this governmentality regime Occidentalism. It is important to note that such 'excess', as Ahıska calls it, the conflicted feelings about these new ideals and/or losses were not publicly voiced or shared at a collective level in this early period, due to the strict authoritarian power of the state. The Occidentalist fantasy, for Ahıska, not only defined how the state wielded its military power, but also described life practices – the way people move, act, feel and think. Ahıska's historical analysis of the bifurcation of reality and its consequent life practices identifies the era when melancholia was forced to bleed into intimate spaces rather than represented to a collective consciousness. In a similar vein, I argue that in the Republic's early period, representing subjectivity and personhood in public was largely precluded by political and social conditions. In other words, the body remained an 'object' of military state power,[11] and wounded personhood could only express itself in intimate spheres where the 'excess' could be contained.

The way people act and move today, however, cannot be understood solely from a perspective which frames contemporary performances by looking at the early Republican era's public life. From the 1940s onward, the political system has changed from a single-party system to a multi-party system, from statist economic politics to liberal and neo-liberal market economy, et cetera. The ideals of modern, secular Turkey have been redefined and reproduced throughout Turkey's contemporary history, in accordance with changing economic, social and political conditions. Through exposure to radio, television, novels and the like, and schooling, army discipline and various other institutional and social practices, people of Turkey have internalized the ideals and 'came to organize their lives, to a certain extent, around the new notion of what being Turkish (and modern) was now about' (Navaro-Yashin, 2002: 12). Ahıska and Mardin's discussions about the ways in which 'people' of Turkey received the 'fairy-tale' fantasies of the early republican era provide us with a rich domain through which we can grasp how 'people' have participated in forming these fantasies and making them 'real'. Thinking through melancholia, I believe, makes us aware of the continuities and discontinuities of the discursive and non-discursive practices of state/society throughout Turkey's history. Thus if we attend to early state attempts to implant national ideals into the psychic domain of the citizenry, and think over the early collective habitus of 'as if' with regards to imaginations of a 'young' national identity, we gain greater insight into the production of historical melancholia as a tactic of military state power as well as a defence strategy by the people.

Taner Akçam (2004) argues that the very ideal of Turkey was based on the feeling of former greatness. He writes that 'Turkish national identity' was imprinted by the long experience of military defeats and the resulting losses of power. In return a nostalgic national ideal emerged that held forth the vision of a glorious past as a future goal. Nationalism, modernization, secularization, and westernization were all imagined to bring forth this former greatness. Judith Butler tells us, 'the historicity of loss is to be found in identification and, hence, in the very forms that attachment is bound to take. "Libido" and "attachment" in such a view could not be conceived as free-floating energies, but as having a historicity that could never fully be recovered' (Butler,

1997: 195). While drawing attention to the historicity of the loss to show that loss cannot be understood as an ontological lack, she also tells us that the historicity of loss can never be fully examined and recognized. In that respect, we cannot argue for and list all possible losses, which give rise to melancholia. Furthermore, insofar as melancholia defines an individual's social and psychic condition, her/his loss never comes to be represented by her/him or through her/him (Freud, 2001). The loss can only be marked in the melancholic's movement or discourses (in the strict sense such as speeches, narratives) (Butler, 2002). Thus, analyzing melancholia in terms of understanding the losses or the history of losses necessarily involves speculative work. However, by describing the work of power on the body and thinking of the body as an institution of melancholia, we assume that the very subjectification process itself generates the loss on the part of the body. For example, Butler considers homosexual love as loss, as that which is lost during the process of becoming a viable social subject.[12]

Here, by following mainly Akçam and Pamuk's insights I argue that the very loss which initiated Turkish political and social subjectivity was the loss of the empire as it was imagined in the early republican period. The 'former greatness'[13] in this context reflects the ideality of the empire, albeit only partly. Nationalism, westernization, secularization, and rationalization can all be interpreted from this perspective as melancholic tactics and strategies which reflect the collective's and individual's incorporation of the ideality of the loss. These formations can also be seen as ideological instruments of power to form melancholic bodies, melancholic collectives, or as Ahıska would have put it, subjects of Occidentalist fantasy. At this point, one may recall Pierre Macherey's statement, 'the state would not be so strong, if its subjects were not so melancholic' (2004: 17). To make these arguments clearer, I will review some recent studies on collective melancholia.

Eng and Kazanjian's collection *Loss: The Politics of Mourning* (2003) addresses the melancholia of collectives that appear as diasporic groups, immigrants in the new world, nations, colonized societies etc. One of the issues that their contributors demonstrate is how melancholia, originally intended to define an individual's psychic and social condition, is also applicable to a collective's psychic and social condition. The loss that initiates a subject also materializes the collective and becomes a condition for a certain sense of community. With regards to nationalism, Eng and Kazanjian write, 'nationalism is itself a formation produced in the space of loss or at least in the space of an anticipated loss. For it is loss that summons the fantasy of the return' (30). Thinking about Irish nationalism, David Lloyd (2003) suggests that, 'particularly state-oriented nationalisms respond to the paralyzing sense of loss by seeking to constitute a new culture and subjecthood around a reinvention of tradition…in the shadows of nationalism, there lurk…melancholy survivals' (217). Butler (2003) points to how, in the face of loss, melancholic communities emerge which 'cannot overcome the loss without losing the very sense of itself as community' (468). The remains of loss are reinvested in the present as the very condition for community and the social. The ideals that constitute the base of nationalism come to the fore as 'rearticulation' of the loss, disguising grief in its (nationalism) relationship with the lost ideal. For Eng and Kazanjian:

> [T]hose remains…are not simple reconstruction, as if they were restored ruins installed in a museum as a record of what was. Rather…melancholia materializes the ghostly remains of an unrealized or idealized potential…indeed, it is precisely by imagining such a space for the remains of the past that those remains can emerge as constricting forces or motivating ideals. (2003: 13)

In that respect, the very ideals of Turkey, its imagined national identity and/or culture might be interpreted as new articulations of the loss which should not be reduced to the empire itself, but needs to be considered as the ideality of empire. The ideality of empire does not necessarily reflect the empire itself, but its unrealized, idealized potentials.[14] Policies and ideologies like nationalism, modernization, and secularization are attempts – which have conflicting and ambiguous aspects – to articulate the remains of the loss. Loss is articulated in each time in different forms (this explains why discourses of ideals are always contradictory or uncertain), any of which are thought to have the potential to restore former greatness or past empirical power. We may surmise, then, that what attached bodies to the new imagined identity was the fantasy of the return of former greatness.

Social categories are not fixed entities. Nor is imagined Turkish identity. Considering the interpellation to Turkishness as the institution of melancholia, should not be understood only as the outcome of the loss of an ideality of empire. Melancholia refers to a psychic and social condition of bodies where various losses (including slights and disappointments, see Freud, 2001) accumulate within the unconscious, in the memory traces of things insofar as the social and political conditions do not allow for the work of mourning. Thinking of melancholia in this way opens up a site where we can comprehend changes in cultural space and public life, and examine such changes in relation to the psychic work of power. In what follows, after arguing for an internal link between melancholia and technology, I will look at Turkey's contemporary cultural landscape and explore the ways in which historical melancholia has become the very base of the social – or the loss of the social – and the very psychic and social condition of the individual, or the loss of the individual.

Technology and melancholia

Cultural studies of technology and historical analyses of technologies in Turkey have generally been neglected in Turkish studies. Although many studies have argued for the significance of technology, techniques and scientific thought to Turkey's cultural ideals, they did not take the argument further. However, encounters with technology both as a meaning-making system and as machines of instrumentality for military battling have been profoundly influential on the early revolutionaries' imaginations for a new nation and in actualizing their dreams of instituting the new country. For instance, the Ottoman Empire's modernization started largely with the technological transformation of the army (Davison, 1990; Tekeli and Ilkin, 1992).

The Ottoman Empire only adopted technologies that would be of help to its army (Davison, 1990). The railroad and the telegraph arrived almost simultaneously in the Ottoman Empire, but the telegraph spread rapidly to all corners of the empire while only a few hundred kilometres of rail were laid (Davison, 1990). Bernard Lewis (1969) contends that Ottomans adopted the telegraph because they assumed it would increase the power of central government. The outcome was different than envisioned; the empire eventually collapsed, and army nationalists used the telegraph when establishing the new Turkey. The telegraph added a new dimension to Ottoman everyday life, one which contributed to wider perceptions of technologies in general. The telegraph has increased awareness of 'time' (time was traditionally split according to pray times in Ottoman society), has emphasized 'speed', especially in decision making, has made recordings of sent and received messages compulsory and has shown that 'technology' can bridge geographical gaps (rural areas can communicate with the Ottoman centre, and the west became nearer as well) (Davison, 1990).

It is significant for my study to understand that notions of 'speed' and 'movement' were at the heart of imagining a modern and progressive Turkey. In its early modernist narratives, a movement towards another time – and the state's role in this achievement – was imagined via 'vehicle' metaphors, mobilizing the nation to an imagined temporal destination. M. Kemal Atatürk (1923) had said that Turkey would 'catch up with the Western civilization'; the 'lethargic mentality of the past centuries' would be abandoned and new standards were to be inaugurated based on the 'speed and movement that define our century' (quoted in Ahıska, 2003: 367). Similarly, the nation's first sociologist, Ziya Gökalp (1917), who profoundly influenced the imaginations of modern Turkey, famously said, 'we shall skip five hundred years and not stand still' (quoted in Ahıska, 2003: 367). According to Gökalp, 'the new civilization was universal and was represented by Western science, technology and intellectual development' (cited in Özbek, 2002: 225).

Images of speed and movement were also contextualized by the east/west divide. These concepts became ideals for the 'nation' in opposition to the Ottoman stasis (Ahıska, 2003). The Ottoman stasis was often characterized as an effect of its ties with eastern values, and of religious dogmas that did not allow for the development of scientific thought or a positivist intellectual tradition. While Europeans moved forward, developed, and enlightened themselves, the Ottoman Empire had stood still: it did not adopt positivist and scientific thinking nor had it been able to produce technology (see Ahıska, 2005). It had, therefore, been defeated and left behind by European powers. The polarity between the east and the west was produced by particular perceptions of time, and the social imaginary in which the modernization project found legitimacy presumed a time lag between Europe and the Ottoman Empire (ibid.). The absence of technology in the Ottoman Empire, in terms of both producing technological machines and generating the scientific basis and technique for their production was seen as temporally backward in the eyes of Republicanists.

Before it became central to contemporary life practices, technology was a metaphor, symbolizing temporal movement, progress and modernity – what was needed to fill the perceived time-lag between the Ottoman and Europe. Technology as a practice was also depicted in different ways by early Republicanists and revolutionaries: on the one hand battles were lost to the Turks because they had 'technology', on the other hand the war of independence was won by the European powers with the help of 'technology'. When Atatürk was asked how he had won the war of independence, he said 'with the power of [the] telegraph' (Davison, 1990). After he and other Republicanists founded the new country, their first accomplishments included material constructions of the technoscape, as we have mentioned above. The majority of work on the ideal of modernity does not take into account technology's role in the formation of national ideals during the early period of the republic and the late Ottoman era. However, it is possible to discern a belief in the might of modernity on the part of Ottoman intellectuals and later Turkish Republicanists, as the utopia of a new country became a guiding vision after encountering western technologies.

Orhan Tekelioğlu writes:

> It was in the area of technology that European powers first demonstrated their superiority, and their might was perceived in terms of consistent military victories over the empire. The first Ottoman institution to feel the impact of the West was the army, where an effort to restructure along European lines not only brought new technologies and new ways of doing battle, but also a scientifically-oriented, naïve positivistic system of thought and the realization of an urgent need to adapt to the requirements of the contemporary world. (1996:198)

Here one can read some of the significant discourses of the early Republicanists and intellectuals, whose faith in the power of modernity grew out of their fascination with technology (which broadly taken, includes scientific thought and positivism). One can surmise, just by looking at these discourses, that technology and techniques made western civilization appear advanced. Göle makes a similar point, that 'technology, rules of conduct, worldview and everything else that makes the West distinctive and sets it apart from more "primitive societies" impart to Western civilization a superiority that lends a presumption of universality to its cultural model' (2002: 84). Technology was the symbol of 'power and authority' as Winner (1993) put it in this landscape, capable of bringing what has been longed for, and replacing what has been lost. If, as Akçam suggests, the ideal was based on a longing for former greatness, and if the discourses of the early Republicanists and intellectuals show us that 'speed' and 'movement' were required to establish the ideal modern nation-state, we can surmise that 'technology' as an imagined category (containing the promises of movement and speed) was always there in the imagination, lodging the missing, the loss, that which has been longed for.

'Technology' as an imagined category also named Ottoman 'guilt'.[15] One of the things the Ottoman Empire lacked, causing its disintegration and backwardness, was 'technology' itself. Technology as guilt, and the desire to catch up with the west in the realm of technology, are inheritances from the Ottoman era. The 'true' technological ideal was the production of 'national technology' which would make Turkey not only compatible with its western counterparts but also more powerful. My suggestions about melancholia and technology continue in the next chapter, where I reflect on the history of 'use' and 'meanings' of technologies in Turkey. I conclude this section with a long quotation from a recent report of TÜBİTAK, a governmental association that promotes the development and improvement of Turkish technology and science. We can trace the 'guilt' taken from the lost Ottoman past, Turkey's historical melancholia and historical technological ambition in this excerpt:

> The countries, such as Turkey, do not have any active role in those processes that carry the seeds of twenty-first century, but they are directly affected by the consequences of them, and they, inevitably, will continue to be influenced deeply. Turkey, in respect of those global processes, has to cope with many problems. Among them, the most vital one is to catch up with technological changes of the age. However, Turkey's challenge has two fronts in this respect. Turkey, which inherited the Ottoman Empire that had missed the evolutionary process towards an industrial society after the British Industrial Revolution, has not surpassed the industrialization threshold yet. Now, while the industrial societies are evolving into information societies, it has to face the problem of keeping up with technological changes leading the new age as well as the problem of overcoming that historical gap. The performance of Turkey in solving these two problems simultaneously will determine her future.

> To cope with these two formidable problems at the same time necessitates gaining ability in science and technology. Gaining ability in science and technology and creating a country that dominates science and technology is the only strategic choice of Turkey. That is the main target in the national science and technology policy. (from TÜBİTAK's report, published as 'The Science and Technology Policy in Turkey' in 1997)

Commodification and institutionalization of melancholia after the 1980s

> It feels like that a shadow of deep consciousness is finally reflected on us. Everybody is about to figure out that this Turkey has no relation with what we have long dreamt of, let alone being the Turkey that we have dreamt. (Etyen Mahcupyan, 2001, Zaman) (my translation)

Many authors agree that after military intervention in the1980s, the state's rigid policies of modernization and secularization started to dissolve. A liberal market economy was adopted, privatization and globalization have become dominant economic and cultural forces, and new cultural and political possibilities – and impossibilities – emerged (Aksoy and Robins, 1997). The government was run in part by civic authorities rather than entirely by the military. New identity politics flourished after the withdrawal of military state power, bringing new political, cultural, and social tensions to the practice of everyday life. With increasing internal migration from rural areas to big cities due to job opportunities, urban spaces have drastically diversified. Neyzi suggests that 'there is an intimate relationship between Turkey's demographic transformation and its identity crises' (2001: 418). With the visible withdrawal of state power from external reality, as Gürbilek (2007) writes, urbanized Turks invented their 'third worlds'. Identifiably different groups like Kurds and Islamists, Arabesque music, and cultural influences brought by newcomers to the city were all reminders to the elites that Turkey's landscape was rapidly changing. As multiple identity politics joined the public sphere, people also came to acknowledge the 'other' in themselves that had long remained silent or had been pushed to the realm of 'inner reality' where excess could be contained (see above) due to the strict modernizing force of the state (Gürbilek, 2007). Along with emergent collective identities like Kurds, Alevis, and Islamists asking for their rights and demanding their share from 'modernity', ultra-Turkish nationalism and neo-Kemalist secularism have also increased, fragmenting social space into enclaves based around identity politics.

Starting in the late 1990s, urban crime rates have risen and terrorist acts have become more violent.[16] Many young people see leaving the country as the only possibility for a better life (Suner, 2001). Stringent broadcasting policies have been partly dismantled and new private television and radio channels created. Television programs are becoming central to (re)making political and cultural life (Navaro-Yasin, 2002). The mediascape has segmented internally, in accordance with various political and social camps, and the discourse of 'othering' is dominant. The language of cheap magazines sets the tone and mode of television programs, and even some news programs cover celebrities' private lives. The historical dream of the European Union is so close and so far at the same time.[17]

This new period is conceptualized as the emergence of 'civil society' by the majority of contemporary theorists, (see for instance Göle, 1994) or is understood only as a transformation of state power from repressive to productive by others (see for instance Navaro-Yashin, 2002). Here, by following Butler's theory, I will develop another way of looking at this period which does not dismiss the truth of the 'civil society' argument that new possibilities for 'collectives' – in the form of identity politics – have indeed materialized after the dissolution of the strict state policies. Nor will my approach ignore the truth of the other argument, that although the state is constantly deconstructed in everyday discourses, it endures as a fantasy and an abstraction to which the people of Turkey have a fetishistic attachment. My argument is that as the military state power withdrew from external reality, historical melancholia became conscious and turned into a cultural institution, which affects people of Turkey in different ways across all aspects of life.

Judith Butler (1997) suggests that melancholia becomes conscious when the state's authority disappears as an external object, and is incorporated into the psyche. Melancholia takes a conscious form only upon the condition of a withdrawal of military state power (176). When melancholia becomes conscious, a conscious representation of melancholia becomes possible. That is to say, as melancholia becomes conscious, the melancholic's ambivalence towards the lost object as both love and hate of the object starts to be represented as oppositional parts of the ego – as a conflict between one part of the ego and the critical agency. Butler writes, 'melancholia produces the possibility for the representation of psychic life' (177). Before melancholia becomes conscious, it remained unconscious; melancholic ambivalence must exist in the unconscious before it can be represented to consciousness. Butler writes, 'the ambivalence that is withdrawn from consciousness remains withdrawn "not until the outcome characteristic of melancholy has set in"' (Freud, 2001: 257 cited in Butler, 1997: 175). The outcome characteristic of melancholy, in Butler's review of Freudian theory, is the psyche split into different and conflicting parts – the ego and critical agency. Melancholia's conscious articulation is intrinsically dependent on this formation. As the psyche splits into parts, melancholic ambivalence permeates consciousness and melancholia becomes conscious. Unconscious forms of love and hate come to be represented through the melancholic's speeches, while the loss always remains unspeakable and unspoken.

By following authors who highlight the dissolution of military state power, and drawing upon the work of critics like Eng and Kazanjian that shows how melancholia defines the conditions of collectives, I suggest that after the last military coup historical melancholia in Turkey has become conscious at collective level. This level of collective consciousness makes multiple forms of representation possible, allowing collectives to access the public sphere, ask for their rights, demonstrate and demand recognition of their differences. While melancholia has encouraged these conditions, it also marks the loss of a social bond as each group creates an enclave in such a way so as not to interact with, or even develop hostility towards, the 'other'. When individual bodies align with collectives, they take others as the object of their negative feelings. To understand how melancholia separates and sticks one to another, we just need to look at the social and political history of the landscape.

Before the 1990s, Turkey experienced three coup d'états, each of which generated social and cultural pains and losses. In 1960, the military took over the government, assassinating (literally) the prime minister and some of the deputies by declaring them traitors. The country was divided into two political camps: supporters of the right wing party and of the left-wing party (the coup was led by the left-wing part of the military and supported by the leftists).[18] In 1971, the military claimed control of the government a second time and sentenced thousands to jail, torturing prisoners and assassinating some young leaders of socialist groups. This further entrenched the polarization between left and right. In 1980, the military appropriated the government and ruled the country in its most violent fashion, by sentencing more people to jail, torturing them, 'accidentally' killing them because of their political orientations or even because they were friends with certain people (see Belge, 2007).[19] The day before the coup, streets in big cities were filled with people demonstrating against the government, the day after these same streets were empty and deathly silent.[20] It is often said that many people had supported the last coup.[21]

Turkey has only recently started to discuss its past, which was filled with military interventions at the public level. The last two coups have largely persecuted the young citizens of Turkey whose ambitions were to 'save the country'. The student movement of the late 1960s and 70s was 'statist' in essence, the left Kemalist-oriented, while the right was more ultra-nationalist. The leftist student movement was not against the Kemalist ideology, its modernization or nationalist ideologies, rather it opposed the governments' 'capitalist' administration policies. Deniz Gezmiş, who was hanged in 1972, says 'You raised me with Kemalist ideas. I grew up listening to memories of the War of Independence. Since then I have hated foreigners. We are the fighters of Turkey's second war of independence' (1972, cited in Neyzi, 2001: 418). As was expected of young Turkish citizens, 'revolutionary' students' 'self' had no value if it was not devoted to the country. Gezmiş says, 'we have made a gift of our lives to the people of Turkey' (ibid.).

The last military coup also marked a new beginning for Turkey. This new epoch is characterized by a collective search for an 'identity', 'individuality' or 'self' which is not devoted to the country, but to itself and to its own good. Within this period, the military past has largely been 'forgotten', and the habitus of acting 'as if' nothing had happened continues to dominate the cultural and political landscape. The new Prime Minister, Turgut Özal, elected in 1983 after the military left the government, promoted a new motto that came to define the 1980s. The motto of the period was 'Turkey is opening to the world', which by and large meant 'consumption' had become the new way to forget the past. The ultimate ambition of an individual was to 'turn the corner' (*'köşeyi dönmek'* in Turkish) by making money quickly and easily (and often illegally), and to consume (Neyzi, 2001). Since opening the country to a liberal market economy and with the new narrations about success and happiness, consumption has become synonymous of happiness and an emblem of success in life (see Navaro-Yashin, 2002).

In this regard, rather than considering the post-coup period as one in which collective mourning took place and 'civil society' finally emerged, we need to see the ways in which melancholia has attained cultural form. Within this new epoch, rather than directing anger or hostility towards the state – i.e. the military – that inflicted losses on bodies, the social body has turned its anger and hostility inwards, fragmenting the social into pieces where a relation has become possible only among similar groups. Cities have become more and more segregated, and in such a way that Islamists and secularists, Kurds and Turks, Alevis and Sunnis, newcomers and long-term residents live in the same city without interacting with each other socially. Rather than becoming a heteroglossia where each corner reveals – and revels in – difference, urban spaces have filled with invisible barriers that block developing relations. When researching public life in the 1990s, Navaro-Yashin conducted interviews with people belonging to different political camps, such as Islamists and secularists, and found that individuals from these groups have no relation with each other at all. For instance a middle-aged woman who has lived in Istanbul her entire life has not ventured to its Islamist districts in the last two decades. When Navaro-Yashin and this interviewee went to Fatih – the famous Islamist district of Istanbul – together, the woman could not disguise her surprise and shock at seeing that her city, Istanbul, is also 'home' to 'others' who do not look like her. She reacted as if these 'others' could only exist and move as a 'sign', as an 'object' of negative feelings, like fear, hate, anxiety (see Ahmed's discussion of emotions in the next chapter), but could not be neighbours, bodies in the same city.

In an interview with Vikki Bell Butler describes how cultural forms of melancholia preclude the ability to 'love' among different groups of a given society by occluding the very means of the social:

> I think that there are cultural forms, culturally instituted forms of melancholia. I suppose by that I would also mean discursively in the broad sense of discourse. It is not a question of this ego not being able to love that person – it's not that model – it's rather what it means to have one's desire formed as it were through cultural norms that dictate in part what will and will not be a love-able object, what will and will not be a legitimate form of love. (Butler, 1999)

In the post-coup period, the possibility of 'love' for the other requires an extra effort, in order to obfuscate any given 'knowledge' about the other. The means by which people are held together have largely been transferred to commodified spaces and practices. As the restlessness that melancholia gives to the body has started to flow into markets, what now binds people to each other are their consumption patterns. Melancholia is institutionalized by channelling the restlessness of melancholic bodies into a domain of superfluous consumer wants, by turning the desire for 'proper interpellation' into the very engine of market-mediated consumption (also see Johnston, 2004). In that way the loss has become covered by the commodity system (Butler, 1999). Commodification commonly reifies certain symbols into markers of identity. Navaro-Yashin argues that consumerism was so central to the social life of this period (the 1980s and 90s) that political conflicts were organized, expressed and mediated through it (79). The state's neo-liberal policies, and the constant recourse to 'individuality' as a condition of liberated collectives concomitantly conditioned people to look for their liberty, individuality and independence in commodified spaces, such as in shopping malls, in the private mediascape (see Göle, 2002). Along with the military state power's withdrawal from the mediascape and technoscape, new private channels in television and radio broadcasting were introduced as mentioned earlier and thereby 'people found their tongue' (Göle, 2002). Consumption of speech has become one of the dominant activities of the period. Both Göle (2002) and Gürbilek (2007) wrote that after the 1980s, people started to think that publicly putting feelings into words would liberate bodies, would provide the means of 'freedom'. As people spoke about their inner worlds, deciphered their sexuality, personal beliefs and dreams, they believed that they became more and more liberated (see Göle, 2002). Göle (2002) argues that the subsequent two decades were based on the collective desire to be rid of the 'shame' that previously would have derived from displaying private lives in front of other people, and a need to shatter the cultural and ethical walls that separate the private from the public. Gürbilek (2007) points out that the entire cultural atmosphere of the 1980s and 1990s has been dominated by the politics of speech. While issues related to consequences of the last military intervention (loss and pain of loss) are forcefully silenced, private lives, personal pains, desires and needs, deprived of their direct politically subversive and resistant content, were put into words in the public sphere (ibid.).

Butler suggests that as melancholia becomes conscious, the melancholic subject comes to believe in the sustaining power of the voice. She writes, 'the melancholic would have said something, if he or she could, but did not' and now the melancholic refuses to speak anything else but herself or himself, leaving a 'refracted trail of what' s/he did not say to or about the lost object. The melancholic therefore is very communicative and loves to speak about him/herself. Freud writes:

> The melancholic does not behave in quite the same way as a person who is crushed by remorse and self-reproach in a normal fashion. Feelings of shame in front of other people…are lacking in the melancholic, or at least they are not prominent in him. One might emphasize the presence in him of an almost opposite trait of insistent communicativeness which finds satisfaction in self-exposure. (2001: 247)

People of Turkey have begun to search for the 'self', the 'subject' or the 'individuality' within themselves, where the sense of the social has been lost. This period has been characterized by a collective search for 'home', and a sense of 'in-betweenness',[22] between lost past and lost future, east and west, has become a part of public life. While the walls of 'shame' were being collectively dismantled to reveal what had been repressed or relegated to the inner world, the ghostly presence of the Ottoman has started to be felt in the public domain, showing the secret space where melancholic bodies can search for their own 'identity'.

Starting from the mid 1990s, people of Turkey started to discover and recover their Ottoman past (see Karaosmanoğlu, 2006). Ottoman food, music, and art entered into public domain insofar as they became commodities, addressing a commodified nostalgia for the lost ethnic and religious plurality, and for the lost Ottoman past. Yet while Ottoman remains were constantly popping up in the consumer market, reminding people that once upon a time ethnically and religiously diverse people peacefully coexisted in this region, the events like the Armenian genocide, forced migrations of non-Muslims, and the early revolutionaries' violent acts persist as taboos which cannot be recognized, accepted or spoken about.[23] The Ottoman was re-produced as both 'self' and as 'other' (Karaosmanoğlu, 2006). In commodified spaces, representations of Ottoman diversity, art, and cuisine were reminders that consumers are not rootless but have an entire history that can provide a means of 'belonging', and yet simultaneously also mark the impossibility of identifying with their Ottoman past by highlighting the differences between the Ottoman and the Turkish social and political landscapes. This ambivalent relation with the lost past was profoundly obvious in commodified spaces (see Karaosmanoğlu, 2006).

Liberty, freedom, movement, speech, identity politics and re-writing history were allowed insofar as they reproduced melancholia in cultural forms and insofar as such mourning was externalized to multiple consumer fetishes. I argue that after the 1980s, historical melancholia has become both conscious and presentable in various public forms, by channelling the potentiality of mourning inherent in melancholia into the world of objects where things cast off their habitual meanings to become vehicles which both sustain the phantasm and are their own mourning cry (Agamben, 1993).[24] In return, social reality has attained more and more phantasmagorical structure. Consumerism is central to this phantasmagoria, not just for purchasing nostalgic goods and services, but also through the cultural forms offered by the media, as something the people of Turkey find compelling even as they hopelessly try to denounce this addiction. Television programs, particularly news and magazine programs which talk about celebrities' private lives, television dramas which are essentially reproductions of early melodramas, and Turkey's version of *Big Brother* shows, have made people addicted to themselves.[25]

As Stiegler writes, consumption intoxicates and disaffects people by turning them into addicts who need consumption in order to 'feel'. He writes, 'consumption becomes a hard drug when one passes from cheerful consumption, which believes in progress, to miserable consumption, where the consumer feels he regresses and suffers from it. At this stage, consumption releases more and more compulsive automatisms, and the consumer becomes dependent on the consumption hit' (2006). One manages and compensates for suffering unbearable loss only by intensifying his/her consumer behaviour, a pattern that becomes pathological. Yet as Stiegler points out, when the people who are supposed to evaluate value turn into a value to be evaluated, when the desire that makes value worth something is liquidated, people come to see themselves as devalued. 'Such is the lot of a society which no longer loves itself', says Stiegler (ibid.). The scene of this devalorizing devaluation is not simply the market, Stiegler argues, it is the hypermarket, the emblem of a hyper-industrial epoch. Yet Turkey, whose economy especially after the 1980s has become dependent on internal and external debts to IMF and other international institutions, is by no means a country of massive industrial production. Turkey's market is instead based on hyper-consumption. Commodified spaces are where society can consume the phantasm of living in a country that has progressed economically, that has finally succeeded.

People are keen to consume goods that correlate with their imagined or ideal life experiences, such as increased wealth and more prestigious social status. The market is organized in a way that people who want to consume but lack the cash, are allowed to purchase goods or services by resorting to their credit cards. This has produced a debt economy, producing new space-time relations, and paradoxically sustaining the belief that people are progressing economically by being more and more in debt (see Ali Ergur's work (2005) on the culture of credit cards in Turkey). The constant feeling of living in debt to tomorrow feeds the desire to extend 'the present' and delimit it with the experiences and imaginations of 'now'.[26] Credit cards work perfectly to sustain the phantasms, which are always in need of more and more fetishes to temporarily cover the loss of loss. In her article on melancholia and fetishism, Comay stresses how these psychic organizations effect a particular relation with time. Both the melancholic and the fetishist tend to freeze time before and after the trauma of loss, fixating on an experience of 'now' that only indirectly relates to the past and future. The organization of commodified spaces works alongside such melancholic and fetishist desire in relation to time.

Freud writes in *Civilization and Its Discontents* that the suffering that arises from our relations with others causes us more pain than any other (2002: 15). The loss of social unity is perhaps the most influential force pushing bodies to maintain their failing and directionless movement of melancholia. In the absence of a social bond, melancholia itself integrates people. People are both connected to and separated from each other by the same affective condition, melancholia. As melancholia is produced and reproduced in the cultural and social sites of everyday life, individuals come to feel that they are part of a larger collective. Institutionalized melancholia interpellates individuals into certain social categories, be it Islamist, newcomers, urbanites, periphery or seculars. Through the force of a culturalized and commodified melancholia, 'Turkey' becomes a felt collective in the sense that what all different people share is the melancholic condition (see Eng and Kazanjian, 2003; Butler, 2003 above).

In the light of these arguments, I contend that cellular telephony, as a global technology used in the 'present' of the global world, as a mobile/personal technology enabling one to experience imaginative movement or departure from where the body is, and as a communication technology whose well-known promise is 'connecting people', speaks to melancholia. As a technological machine, given that 'machine' means power and authority, as a purchasable object given that consumption has become the primary collective experience, as a recording device for ordinary autobiographies, as a telephone book which reveals how lonely or popular one is, and as a tool which can only work for its 'master', the cell phone speaks to and through the imaginations of everyday life in Turkey. With these multiple trajectories in mind, I try to grasp how cellular telephony works with melancholic imaginations and affectations in Turkey. I continue with an analysis of how historical melancholia, as a psychic and social condition of both the body and the collective in Turkey, is reflected in the very performances of the technoscape, and how the technoscape holds the promise for overcoming melancholia in Turkey.

3

RETHINKING THE TECHNOSCAPE AND CONTEXTUALIZING CELLULAR TELEPHONY IN TURKEY

This chapter stresses how relational concepts such as imagination, attachment and emotion define inherent dimensions of social experiences of the technoscape. It proposes to consider the technoscape neither as merely a container of technologies, techniques, bodies and capital, nor as just a fluid landscape of culture where the local and the global are in constant negotiation. Rather it argues for an analytical sensitivity when considering the technoscape – and its attendant imaginational and emotional experiences – as both a global and a local site for social practices. After providing historical and introductory accounts of key dynamics and issues with regards to 'technologies' in Turkey, I will conclude by demonstrating why the technoscape in Turkey should be conceptualized as an historical dream of presence, outside of which cellular telephony cannot be effectively and affectively thought.

Movement and imagination

Movement and imagination are two key concepts, describing the promises, potentialities and experiences of the technoscape. Arjun Appadurai (1990) first used the term 'technoscape', along with other 'scapes' like ethnoscape, finanscape, ideoscape and mediascape, to describe the fluidity of globalization. His account of globalization is based on an argument that something has changed in the world; that 'the imagination has become a collective and social fact' as a result of the technological changes over the past century (1990: 5). For him, electronic technologies' mediation of everyday life has turned imagination into global social practice. This in return has become the basis for the 'plurality of imagined worlds' that characterizes the global era (ibid.).

Each scape designates different dimensions of global cultural flows that together constitute building blocks of the 'imagined worlds' of people around the globe. The suffix-scape denotes the fluidity and irregularity of terrains that 'are deeply perspectival constructs, inflected by the historical nation-states, multinationals, diasporic communities, as well as subnational groupings and movements and even intimate face-to-face groups, such as villages, neighbourhoods, and families' (1990: 5). Appadurai's technoscape refers to the global configuration of technology moving at high speeds across national boundaries and around the globe. His description of this flexible and ever-shifting landscape leaves the impression of a shapeless container comprised of technologies, techniques, labour, and capital that enables movement of things, data, information and bodies and thereby provides new resources and means for imagination.

Following Appadurai, the term now circulates in recent work, ranging from globalization studies and transportation research to urban, media and cultural studies. The landscape (as a space for experience and imagination) is often considered as a kind of technoscape in which movement of bodies and things are largely organized and regulated by technologies and technological configurations (see Harries, 1992; Sheller and Urry, 2006a; 2006b). Physical spaces are increasingly seen in terms of flow, movement and of flux by many authors (Urry, 2003; Clark, 2001; Graham and Marvin, 2001). Sheller and Urry (2006a) conceptualize contemporary experiences of space as multiple engagements with the technoscape that occur 'whenever people ride in a train, make a phone call, read a computer screen, simply step off a pavement to cross a road, or hike one marked trail. Data, pictures and sounds also flow through technoscapes' (8). Importantly, Sheller and Urry (2006a) highlight the technoscape's association with specific equipment, how 'a car, a mobile phone, a camera, a screen, a hiking boot and so on' can enable a particular production and consumption of space (ibid.).

In all these works, imagination and movement remain core concepts for defining the character of the technoscape. The technoscape refers to both mechanical and informational technologies which facilitate 'movement' or 'action' that is not limited by the physical deterritorialization of the body. Movement is not restricted to images, data, information, but also the virtual and imaginative movement of the body. By enhancing the possibilities of movement, the technoscape also augments the means of imagination and transforms it into a daily, ordinary and collective practice. Imagination as a bodily practice already implies a shift towards a certain exteriority from the present, from action and choice not previously determined by causal laws (see Sartre, 2001). The terms 'imagination' and 'imagery' indicate a capacity to visualize that which is not the case (ibid.). Through generating new practices in time-space, the technoscape drastically changes traditional understandings of distance and proximity, absence and presence. For instance, by providing the means of virtual movement, digital technologies have transformed what we think of as closeness, creating a feeling of co-presence while bodies remain physically distant (Licoppe, 2004; Urry, 2002).

The whole promise of the technoscape derives from its capacity for displacement and circulation. The boundaries between absence and presence, distance and proximity are blurred by technologies, letting us believe (or hallucinate) that we can all experience movement. Capitalism is also reinvented, as places are re-spatialized, and market conditions are radically altered (Goankar and Povinelli, 2003) in such a way that new forms of global asymmetries take place (which will become clearer below). Likewise, 'distances' are recreated, and 'displacement' of some others (like immigrants and refugees, the ones who do not have 'proper' passports for free deterritorialization) is more controlled under the surveillance of new technologies.

Attachment and movement

In the 'culture of circulation', as an effect of globalization (Lee and LiPuma, 2002), mobility becomes a sign of liberty and modernity. In an era of globality,[1] bodies become more attached to 'movement' as an ideal and as a requirement of the global world (Ahmed, 2004b). However, 'movement' becomes an ideal on the condition that not all bodies nor all objects and images have an equal relationship to movement (see Albertsen and Diken, 2001; Skeggs, 2004; Beckmann, 2004). Globality depends on the movement and circulation of some bodies, images and objects and not others (Ahmed, 2004b). For those who are able, an attachment to movement becomes a new form of social bonding (ibid.). This attachment links one to others, forming a 'global community' which consists of those who can move – virtually and physically – in contrast to others who cannot (see Nussbaum, 1996).

The technoscape, which comprises the means of physical, virtual, and imaginative movement, plays a significant role in this attachment to movement. Its velocity of information and its global reach has made bodies too aware of their slowness, fragility and their vulnerability to time and space. Such perceptions affect collectives differently; some feel more vulnerable to time and space than others, some feel like they are living in slower zones of 'flow' and 'movement' than others. The quality of the technoscape, the degree of its absence or presence in an era when global cultural forces prevail, marks differences between regions, countries, or even collectives. As Doreen Massey (1993) argues, power geometries play a critical role in the perception of technoscapes, so that space-time relations vary across different geographies. She writes:

> [D]ifferent social groups and different individuals are placed in very distinct ways in relation to the flows and interconnections. This point concerns not merely the issue of who moves and who doesn't…it is also about power relation to the flows and the movement. Different social groups have distinct relationships to this anyway – differentiated mobility. Some are more common in

> charge of it than others; some initiate flows and movement, others don't. Some are more on the receiving end of it than others; some are effectively imprisoned by it...There is the dimension of the degree of movement and communication, but also dimensions of control and initiation. (61– 62)

Differentiated mobility is integral to the social construction of space and time in an age when imagination is available to all, but the very experience of movement (virtual and physical) varies among different groups, collectives and nations. As Massey stresses, the issue is not just if one has access to the flow or movement but also how people engage with the means of mobility. Many recent studies show that in particular regions of the world, the mode of engagement with the technoscape determines the sense of and the feeling about the 'landscape' in which people live (see Chakravarty, 2004; Rafael, 2003; Bull, 2006; Lemish and Cohen, 2005). Larkin (2004) demonstrates how the use of second-hand videos, as the only possible film-viewing practice in Nigeria, generates among Nigerians the sense that Nigeria lives outside of 'history'. The inability to adopt due to the institutional or structural inaccessibility of technological novelties translates into a collective feeling of resentment. Conversely, the ability to adopt the newest technologies can engender a collective pride, such as in the Philippines (see Rafael, 2003).

The technoscape is therefore a site where different people come to imagine and feel together, yet it also demarcates the very distances between different geographies by inserting new boundaries and barriers. The technoscape connects, attaches one to the other, and frames kinship as different people are brought together by the same social practice. However, it also separates one from another on the basis of language differences, class differences, and gender disparities. These diverse issues tell us that the technoscape is not only a collection of materialities that produce collective imaginations but also a plane of affect[2] where social attachments and detachments affectively take place, where bodies and collectives affectively take shape.

As Callon and Law comment, 'emotions, to be affected, all have to do with travel, with circulation. To be moved, to be transported, the trip, these are metaphors for displacement' (2004: 10). Just as 'movement' is inherent in the practice of imagination, the relationship between movement and attachment is implicit in the concept of emotion. Emotion implies a certain movement in the sense that emotions are 'what move us' (Ahmed, 2004a, 2004b). The word emotion, *emovere* in Latin, refers to 'move, to move out' (ibid.). So 'what moves us, what makes us feel, is also that which holds us in place, or gives us a dwelling place' (Ahmed 2004a: 27). Emotions are also about attachments, about what connects us to this or to that. So 'what attaches, what connects us to this or that place, or to this or to that other' is also 'what moves us, or what affects us such that we are no longer in the same place' (ibid.). Therefore movement does not take the body away from where it is, but essentially connects it to other bodies. In other words, 'attachment takes place through movement, through being moved by the proximity of others' (ibid.).

Emotion and imagination: affectivity of the technoscape

The issue of the affectivity of the technoscape is critical to a social study of technology, and it can be investigated through different scales ranging from the feeling of the individual body with the technological tool to the regional and national technoscape. When trying to grasp why some technologies are particularly popular, why and how they become objects of collective attachments, and even of addictions, we need to take into account the affective possibilities and experiences of these technologies. For both the individual and the collective, from their physiological to sociological aspects, technologies necessarily speak to the world of emotions, affective relations. To make it clear what I mean by the affectivity of the technoscape and what affectivity I will study in this book, I need to comment on and clarify the relational notions of feeling, affect and emotion.

These three concepts are widely used in cultural and social theory, however there is no consensus about how they are differentiated or about how they should be defined. Theorists have oscillated between conceptualizing emotions as either 'inherent' or 'socially constructed' (see Lupton, 1998; Williams, 2001). Psychological works often consider emotions as the 'inside getting out', whereas early sociological works conceived of emotions as the movements of feeling that do not originate in individual consciousness, but come from 'without' (Durkheim 1966: 4 cited in Ahmed, 2004b: 28). From the latter perspective, the individual is not the origin of feeling; feeling comes from outside (the social world), and moves inwards to become an integral part of our being. In contrast to these two positions, Sara Ahmed argues that 'emotions work to create the very distinction between the inside and the outside, and that this separation takes place through the very movement engendered by responding to others and objects' (2004b: 28).

For Ahmed, emotions occur when the flow of sensations and feelings becomes conscious. Emotions come with the recognition of feeling (for instance, feeling pain) and judgment about it (it's bad) and finally lead to action (move away). Feelings are 'not about the inside getting out or the outside getting in, but that they "affect" the very distinction of inside and outside in the first place' (ibid.). By recognizing and interpreting sensations – responses to impressions of objects and others – and transforming such sensations into emotions and judgments, bodies and worlds materialize and take shape.

Emotion and affect are generally differentiated in both sociological and philosophical works. While emotions are described as more cognitive experiences of the body, affect refers to non-cognitive, almost unconscious, 'irreducibly bodily and autonomic' intensifications (Massumi, 2002: 28). We feel the physiology of being affected and then emotionally respond through the resources of our memory. Emotion then 'represents the assemblage of any affect with our previous experience of that affect' (Nathanson, 1996: 13). Affect emerges out of transition (relations, encounters) between bodies, between body and space or between body and thing (Anderson, 2006: 736). Affect is contextual, situational, trans-situational (Massumi,

2002), and always involves 'potentiality' and 'excess' (Anderson, 2006). Being affected and affecting are two aspects of bodily movement that point to the openness of the body to the other, the world or the object (Massumi 2002; Anderson, 2006; Ahmed, 2004). However, in contrast to emotions, affect affords a potentiality and an excess which can transcend the boundaries of the social (Anderson, 2006). Affect can shake up the body, may realize the potentialities that are described by bodily intensities and might outrun the cognitive limit of the body. The potentiality and excess of affect is precisely what might threaten the social, political, historical foreclosures that make up the subject or the social body (see Probyn, 2004; Anderson, 2006).

Although I agree with Ahmed (2004a, 2004b), when she argues that offering discrete definitions for each of these three concepts might lead us to neglect their historical mediations and interrelations, it is important to note that affective relations do not only point to the closing down of the possibilities for the body that is captured by the social but also to the potentialities of transcending the social to a certain extent. Thus the very distinction between emotion and affect needs to be underlined: not all affective relations are emotional, nor are all bodily attachments to the world of objects. In the case of bodily attachment to technological objects, this attachment is characterized by the artefact's affective potentialities, its sensory relations with the body (see Verbeek, 2005). For this reason, defining affect as 'potentiality' and 'excess' that links us to the very notion of 'hope', to the hope of arrival where the body feels 'at home' is crucial (see Anderson's work, 2006). Simondon (1989), who worked on technologies, writes that 'affectivity indicates and comprises this relation between the individualized being and preindividualized reality: it is this to a certain extent heterogeneous in relation to individualized reality, and appears to bring it something from the exterior, indicating to the individualized being that it is not a complete and closed set (ensemble) of reality' (cited in Hansen, 2004: 125). In other words, affectivity resonates with the possibility of being beyond identity, beyond the social closures of the body, beyond collective feelings and shared emotions.

The promise of technology exists in direct relationship with the affective possibilities of living in reality. Technologies alter the very basis of our sensory experience and drastically affect what it means to live as embodied human agents (see Hansen, 2004; Hayles, 2002). Experiences with technologies are also grounded in the biological potential of human beings. For instance, the source of virtual reality is not technological, but rather a biologically-grounded adaptation to newly acquired technological extensions provided by new media (Hansen, 2004). Yet what is physiological is also psychological and social; the source of meaning in technology is the affective body itself. In his theory of techniques of the body, Marcel Mauss (1973) tells us that the body, the 'total man' and his actions need to be understood as a triple instead of a single consideration. This tripartite approach requires one to consider the body as a physiological, psychological and a social instrument. In my view, the concepts of affect, feeling and emotion, with their physical, psychological and social dimensions, are highly useful to understand how the body relates to the technoscape and to technologies.

Affectivity introduces the power of creativity and possibility into the life of bodies, in contrast to socially constructed structures, ideas, representations and identities. The affectivity of technologies transfers affective 'power' from the object to the body (Levy, 1997). Technologies become promises of 'exit' and 'arrival' by triggering the affective potentialities of the body. Technology becomes a call for the experience of the imagined as it triggers the potentiality of the feeling body that essentially feels out-of-space. A technological tool is transformed to a technology insofar as it accommodates the gap between the lived and the imaginable lived (see Hansen 2004; Wills, 1995). Thus, the affectivity of the technoscape is not only bound up in historically and socially shaped emotions and feelings, but also promises to actualize affective potentialities. Although affects are recognized, judged and interpreted according to what we already know, their 'potentiality' and 'excess' still linger even as technologies organize the bodily practices of everyday life. In this respect, a technology's 'affectivity' can be considered as both what the tool promises in relation to bodily sensations, as well as feelings that can alter the product of social structures – such as social identities – in which emotions and unconscious feelings have become the second nature.

Thus, the affectivity of technology needs to be studied alongside an historical understanding of the technoscape (i.e. how technologies have been imagined, idealized, experienced and sensed in a particular society). Perceived and experienced affective potentialities of technology 'can only be comprehensible within that specific cultural frame of meaning and style and larger historical frames of power and discipline' (Appadurai, 1990: 148). As Grossberg (1990) puts it, 'the same object, with the same meaning, giving the same pleasure is very different in different affective contexts. Or perhaps it is more accurate to say that different affective contexts inflect meanings and pleasures in very different ways' (80).

Affective relations also influence how bodies position themselves in social space. Ahmed's conceptualization of emotions explores this aspect of affective attachments. She suggests that the emotions which align individuals with collectives always take some 'others' as objects of negative feeling; in the presence of these others (objects of fear, hate), the collective surfaces and materializes. Hence, collectives are largely constituted on the basis of how they feel for 'others' rather than how they feel for the collective. Emotions attach one to another, to a particular object, or to a particular practice always by excluding others. In Ahmed's work the technoscape, as a site of globality, necessarily has to take 'some' as others who cannot be admitted into the global body, for these ones remain too attached to the particular, the local, or the home. Those who cannot move away from home and from local attachments become objects of a 'feeling' that aligns individuals with collectives. Ahmed suggests that fixing some as others is precisely how globality 'surfaces' as a felt collective. In her analysis of the website of 'global nomads', Ahmed (2004a) shows us that what constitutes the collective of 'global nomads' is their feeling for the ones who could not become nomads. The Internet, which brings many scattered agents around the world into contact, becomes a perfect match for nomads who want to share their pride with other nomads in the presence of such others.

Many studies show that some individuals, collectives and societies are more passionately attached to certain technologies than others. Cellular phone technology was adopted more quickly by the 'developing countries' than by the 'developed countries' (see Sheller, 2004; Katz, 2002). Some scholars see the popularity of cellular phone use in the 'developing world' as a process of 'globalization from below' driven by cell phone users (see Eickelman, 2004; Alhassan, 2004; Nyiri, 2005). These works draw attention to the desire of citizens in the 'developing world' or the Third World to become part of the 'global community'. Not wanting to be excluded from the 'networked' world (see Castells, 2000) and from 'global history' conditions their determination to be part of the global technoscape. The emotive force that moves people toward participation in the virtual community is also a shared fear of being late, remaining too local and being like these others that are left behind, left outside of history or of the civilized present. Either as immigrants in the First World or as the citizens of the Third World, the marginalized demand a place in the technoscape and access to the affectations that it promises (such as subjectivity and agency as conditions for social recognition[3]). Like the car has been for many countries (Ross, 1995), the cell phone works in the social imaginary as a symbol of national economic virility and modernity (Pajnik and Lesjak-Tusek, 2002; Zhang and Harwood, 2004). For the non-west the transportation and telecommunications industries signify development, progress, liberty and modernity (see Parla, 2003; Rafael, 2003).

Hence, once a nation's ideals are aligned with the concept of 'progress' that is always already historically connected to models of western modernity, adopting western technologies comes to function as a 'sign' of social progress. Similarly, once the concept of 'liberty' becomes associated with 'mobility', the global technoscape can provoke a collective disillusionment, as the possibility of experiencing virtual and imaginative mobility brings more liberty to the ones whose movements are largely controlled and prevented. In social spaces where the social is fragmented into enclaves of difference an engagement with the technoscape can provide an illusory sense of 'social mobility'. At this point, one may recall Fischer's argument about the ideology of the information society: 'ideologies function to blur and conceal important distinctions. The idea of information society as ideology, serves to conflate techno/logical advance with social progress' (cited in Winner, 1997: 12-13). For countries that export technologies, the social reality that sustains the ideology of the information society is the fantasy itself in which using and adopting technology feels like possessing the 'power and authority' of technology (see Winner, 1997).

Local engagements with the global technoscape become an important part of how nations, regions and communities negotiate larger cultural forces in the contemporary world (Goggin, 2006: 42). When localizing a global technology, this domestication of the global technoscape necessarily involves the production of affective strategies that engender the social practice of a particular technology as a space for negotiating differences and similarities with the globally networked world. It is important to stress that the process of domestication works for and against globalization. The domesticated technoscape unfolds as a hybrid of global and local spaces in which the 'local' use of technologies functions as a process of participation in and resistance to globalization (Pertierra, 2005). Affective appropriations of the technoscape play formative roles in this process, even as domesticating a technology may very well be discussed as one of the strategies of globalization, which deals with 'differences' by commodifying them (see Hall, 1991; Edensor, 2002; 2004). Such commodification lets global capital operate in local markets effectively, however it also contributes to collective imaginations, desires, dreams and feelings. As Appadurai notes, consumption is not outside of the very sphere of imagination, affectivity and of bodily desires (1990).

A particular technology or technological artefact becomes localized in social practices by making connections to what is assumed as local. How a particular technology responds to historical imaginations, idealizations, and feelings; what kind of affective potentialities it suggests a body might experience in relation to historically and socially constituted emotions; how it promises to alter the current space-time relations of the collective or of the individual; how it can be used in social struggles, for positioning oneself within social space; and how it connects and separates people are all integral to the process of localizing a global technology. Producing particular performances of technologies involves opening up new possibilities for bodily feelings and feeling for how the local or national technoscape relates to the global technoscape and the 'others' that are part of it.

Just like any technology, the performance of cellular telephony is mediated and largely determined by historical imaginations, idealizations and collective feelings. That is to say, a cell phone, with its standard features and capabilities is only a thing, a tool or a gadget. The cell phone becomes a technology when it is imitated; as it is socially and culturally appropriated; as it is affectively domesticated (for countries that do not produce but export it); as it serves to collective imaginations, idealizations and dreams; and as it promises an exit from the historically constituted foreclosures that limit the possibilities of the body. How a technology is used, to what extent it becomes a social practice for collectives, and how it becomes an object of collective attachment or addiction are largely determined by social, cultural, affective, economic and political conditions. For countries that export technologies, the domestication of a particular technology entails a series of appropriation strategies and techniques that speak to historical imaginations, collective ideals, dreams and patterns of affective relations and dominant collective feelings.

To work on the historically contingent organizations of local technoscapes, we need a conceptual approach that reflects the specific cultural, historical, political, and social conditions within which technologies are effectively and affectively domesticated and imitated. Jonathan Sterne (2003b) recommends working with Pierre Bourdieu's social theory in order to grasp how history and social structures come into play in transforming a tool into a technology that is socially practiced and produced.

Habitus and social space

Bourdieu's social theory encourages us to look at how and why the social enters into our bodies. This includes analyzing how and why historical, structural, and objective realities determine the way we think, feel and act, whether we are conscious of what we are doing or not. His concept of habitus designates how social structures are embodied in everyday practice. Habitus refers to understanding bodily practice as one in which social structures (e.g. ideas or representations of abstractions like class and gender) are incorporated by agents and 'delimit how they can move and in which spaces they can move' (Probyn, 2004: 336). The habitus is formed and reformed through interactions with material and social worlds. It is not simply the reproduction of social structures; individuals develop habitus through processes of socialization and classification, and are constantly changing based on their interactions with social structures.

Sterne (2003b) suggests that 'technologies are subjects of habitus, they are organized forms of movement' (370). Technologies involve embodied social knowledge and the body acts with them (technologies), most of the time by not being distinctly alerted about their presence. Although each technology refers to different bodily practices and social dispositions, technologies are in general integral to the habitus in the sense that they generate practices, are made up by these practices, and become means for positioning oneself in the world. The habitus refers both to incorporating social structures like a 'genetic knowledge', and to the spontaneity which enables agents to operate within social space according to 'one's position in a field and one's access to and possession of certain kinds of capital resources' (Sterne, 2003b: 375). Capital is integral for maintaining distinctions between classes and collectives in social space. While economic capital refers to money and property, cultural capital defines cultivated dispositions, such as aesthetic values, academic credentials, and material objects that require specific knowledge to appropriate them materially and symbolically. Closely related to economic and cultural capital, symbolic capital is a means of socially recognized legitimization which 'works partly through the control of other people's bodies and belief that is given collectively recognized capacity to act in various ways on deep-rooted linguistic and muscular patterns of behaviour, either by neutralizing them or reactivating them to function mimetically' (Bourdieu, 1990: 69). Finally, social capital functions to show one's networks of family, friends and contacts. Each form of capital works to establish one's position in social space.

Agents are located in social space 'according to their position in statistical distributions based on the two principles of differentiation...economic capital and cultural capital' (Bourdieu, 1990: 6). In other words, social space is where differentiation takes place based on 'a set of distinct and coexisting positions which are exterior to one another and which are defined in relation to one another through their mutual exteriority and their relations of proximity, vicinity, or distance as well as through relations of order, such as above, below and between' (ibid.). The main idea of being in social space is to be different from others. The position one occupies, by possessing different kinds of capital, is integral to the struggles to conserve or transform representations and experiences of social space (ibid. 12). Crucially, while social structure regulates position, one's position is not fixed; there is always room for spontaneity, improvisation and creative social action. Yet, spontaneity and improvisation are bounded by cultural memory, by history. Potentialities are inscribed in dispositional relationships, ones of body and of situations (Sterne, 2003b). The struggle to realize the potentialities and exceed the historical boundaries is what gives character to social space and practice.

Technologies occupy a significant place in social positioning and struggle. Sterne writes, 'a technology is always at any given moment, socially located. It is always implicated in social struggle' (2003b: 383). As embodied practices and also as possessions, technologies may serve to distinguish one group from another and to produce commonalities between different people or groups in the same social space.

A technology may well function as cultural and symbolic capital, becoming a form of agency, prestige and control. Bourdieu's conception of habitus – capital and taste – is essentially based on categorizations of different classes, groups or collectives. The question of how technology comes into play in the social struggle of different groups can only be answered by looking at how particular performances are organized through technologies, and to their basis in historical embodied knowledge, collective feelings and emotions.

The force of history is played out in the ways that agents position themselves and through their daily practices or ways of being in the world. Bourdieu (1990) writes:

> In practice, it is the habitus, history turned into nature, i. e. denied as such, which accomplished practically the relating of these two systems of relations, in and through the production of practice. The 'unconscious' is never anything other than the forgetting of history which history itself produced by incorporating the objective structures it produces in the second natures of habitus. (78)

Habitus and social space are useful terms for thinking through technologies and the technoscape. Bourdieu's theory promotes the investigation of how agents embody social structure and make it their own. This makes possible an analysis of how historical imaginations, idealizations and beliefs are embodied and how they pervade agents' practices. In an effort to understand why and how some technologies become objects of collective addiction, the concept of habitus can powerfully link a particular technological performance to its historical background, inevitably exposing collective ideals, feelings and dreams. Although Bourdieu does not explicitly describe how idealizations and imaginations contribute to the habitus, this concept allows us to think of them as part of embodied knowledge and belief. Appadurai suggests that the connection between the imagination and social life in the global era clearly retains some of the force of Bourdieu's idea of habitus, but he mentions 'the stress must be put on his idea of improvisation, for improvisation no longer occurs within a relatively bounded set of thinkable postures but is always skidding and taking off, powered by the imagined vistas of mass-mediated master narratives' (1990: 55-56).

For Bourdieu, emotions are part of embodied belief. They are also social and become a sort of second nature that replaces the body 'in an overall posture which recalls the associated thoughts and feelings' (Bourdieu, 1990: 69). Probable and improbable movements and practices that make up habitus (Sterne, 2003b: 381) also involve emotions, according to Bourdieu. As Probyn writes, 'emotion projects the habitus' tendency to continually frame and adjust between the unlikely (possibility) and the likely (probability)' (2004: 336). Emotion, in Bourdieu's theory, 'seems to be the body's way of registering its return to "the present of the presumed world the only one it can ever know"' (Bourdieu, 1990: 65 cited in Probyn, ibid.).

While emotions appear to finalize the habitus by being a sort of cognitive adjustment mechanism, the physicality of the feeling body and the potentiality of affect seem to be left aside in Bourdieu's approach. Probyn argues that 'bodies continually try to escape that finality' (344) and the body is also its reactions 'to the dryness, the light, the history, which are enfolded fully within that particular habitus' (345). In Bourdieu, practice is conceptualized in relation to physicality, yet being outside of the social, the physicality of the habitus as feeling body remains largely under-analyzed. Probyn (2004) suggests that the habitus is also where the body calls out its hopes and discomforts because it feels 'out of place'. If the habitus refers to spontaneity, improvisation and creative action as well as to socially and politically regularized practices, its conception needs to make room for the potentiality of affect, and for affective attachments which do not close down the possibilities for the body but open up potentialities of hope.

The promise of 'exit' and 'arrival' is inherent in all technologies (see below). Technology, in that sense, necessarily speaks to the imagined potentialities of 'hope'. So if we are to understand technologies as subjects of habitus, we need to account for how the affectivities of technologies contribute to the body's experience of imagined 'travel' and 'movement'. If technological habitus involves a routine of bodily practices of technology, what transforms the practice into a routine is precisely conditioned by the 'hope' of arrival that is imagined as a probability of that practice. To understand the particular habitus in relation to a particular technology, we need to take into consideration imaginations and idealizations, embedded in historical knowledge and belief. Now, I will try to provide historical accounts of the key dynamics and issues that have been effective in shaping the perception, imagination and experience of technologies in Turkey.

Technoscape in Turkey

Since Turkey first started dreaming of modernity, notions of technology have been integral. While there are numerous studies of Turkish modernity, only a few focus on technology. Thus our historical analysis of the technoscape in Turkey is tied to a search for hints that can reveal how the technoscape was envisioned in Turkey throughout its modern history. One area that opens up a relatively rich site to grasp how the early technoscape was imagined and experienced on a daily level during the transition from Ottoman to Turkish Republic is canonic Turkish literature.

Jale Parla (2003), a renowned literary critic, suggests that Turkish novelists have been preoccupied with cars (along with other vehicles such as carriages, buses and tractors) from the very beginning, which fostered a subgenre of car novels. Parla argues that in all these car novels, the vehicle appears as trope rather than as a means for transportation (2003: 535). She writes, 'the trope of car becomes important at least two ways: in its relationship to the machine and in its signification of a particular kind of space that has become meaningful in Turkish modernization' (ibid.). The first car novel in Turkish literature was *Araba Sevdası* (*The Carriage Affair*) written by Recaizade Mahmut Ekrem, plotted and written in the Tanzimat period (the early modernization period in the Ottoman Empire).

The lead character, Bihruz, is a typical westernized dandy whose one joy in life is his carriage – a sign of his excessive devotion to western culture. The novel criticizes the over-westernization of 'züppe' (beaux) of the time, by labelling such admiration as evil and sinful (Gürbilek, 2003). Nurdan Gürbilek suggests that the car in *The Carriage Affair* is 'the symbol of not only changing places but also changing identities. It represents the promise of a second life, the effort of being someone other than one's inadequate self, to imagine and show oneself as the other, the attempt to close the distance between traditional Süleymaniye and modern Çamlıca[4] (or between belated modern Çamlıca and Paris itself)' (613). By the end of the novel, the carriage, or the promise of being someone else, 'turns out to be the very symbol of a modern technique that cannot be mastered, the symbol of the foreign toy' that leads one to an inevitable accident[5] (ibid.). Throughout Turkey's modern history, car novels use the vehicle's in-betweenness or liminality to depict a dilemma: 'the desire to be the other and the fear of losing oneself in the other' (ibid.). Parla writes:

> As the car stories are plotted in different periods of the Turkish novel by different novelists, these stories which begin with the seemingly innocent acquisition of cars, grow into enigmatic narratives of possession and dispossession, empowerment and loss of power, function and dysfunction, maturation and infantilism, narcissism and fetishism, fragmentation and self-destruction, not to mention a whole century of estrangement and a feeling of inferiority inspired by the conflict with the west. (2003: 536)

Alongside vehicles, the clock was a recurring technological sign in Turkish novels, and symbolized an eagerness not only to fill the gap of an ontologically lacking and incomplete self, but also the incomplete and wounded subject of the modernization of state power (Parla, 2003). In the transition period from the ethnically diverse, disintegrating empire to the modern republican nation-state, encounters with technologies involved interacting with the western powers. In the late Ottoman period, large cities were filled with post offices. Those that were working properly, enabling safe and fast communication, were the European ones in the Ottoman territory, while the Ottoman postal system was disorganized and slow; one of the first things the new republic modernized was the postal system (Demir, 2005). Cars, carriages, clocks, and watches had all been imported from western countries. The technoscape was truly a foreign land.

After the Turkish Republic was founded in 1923, in 1927 the first attempt to build a radio broadcasting system took place as a requirement of the national ideal of 'catching up with the western civilization' (see chapter two where I discussed this motto of the early republican era). The radio was not only a symbol of modernity, but also a useful tool for nation building. In the previous chapter, I discussed Meltem Ahıska's work on early radio broadcasting as it demonstrates how the people of Turkey deliberately maintained the national fantasy of modernity. Here as we focus on the technoscape, we shall add a few more things to what we have said earlier. In her analysis Ahıska (2005) shows us that although radio entered many urban homes as a piece of furniture, it remained a 'foreign' houseguest. She writes that the performance of 'as if' was actually aimed at the imagined western gaze. The radio was not seen as a medium that would reach 'Turkish audiences'. Rather, according to Ahıska, it was thought of as a connection to the west. The whole performance of radio broadcasting was thus geared for the radio – for the imagined western gaze – rather than through the radio (ibid.). Radio technology was essentially 'foreign,' something that the people of Turkey had to welcome just like a respected guest, but with which they never interacted intimately. The radio was simultaneously an embodiment of the imagined western gaze and the representative of the state, and both were still strangers to the audience.

The radio also became the army's vehicle for announcing coup d'etats. Turkey had experienced three military interventions in 30 years (in 1960, in 1971 and finally in 1980) and its people found out about at least two of them through radio broadcasting. Between the 1950s and 1980s, the radio was largely a medium for news. 'Listening to the agency' was the typical idiom to describe the social practice of listening to the radio. If technologies obtain meanings, functions and values through the social practices they engender, radio technology was characterized by state violation and manipulation, up until strict modernization and nationalization policies were partly dissolved in the 1980s.

Radio's perceived strangeness did not derive only from its maternal connection to the state power, but also from how 'technology' in a general sense was imagined as a foreign land. This, however, should not be understood as proof of the 'inevitable' distinction between society and technology or between body and machine. I am trying to designate the cultural, historical and social barriers between society and technology, body and machine that were created by national imaginations and aspirations. Technology was a strange land in the social imaginary because it was imagined and experienced as such. Its imagined social and historical alienage mediated the very bodily practice of technologies, so that hesitancy, anxiety and fear characterized bodily relationships to technologies for a long time. For instance, household appliances, such as pressure cookers, washing machines, vacuum cleaners, and water heaters were appealing both as symbols of modern domestic spheres and for the speed and comfort they brought to domestic life, yet they also spread anxiety and fear. Murathan Mungan, a well-known Turkish poet, writes, 'we were so used to hear the news of explosions of pressure cookers during my childhood' (2000). Once people became familiar with using pressure cookers and water heaters, those that ran on gas started to explode and kill family members. In addition, Turkey has always been a land of frequent traffic accidents, killing hundreds of people every year (see Gürbilek, 2003; 2001). Mungan argues that because we are used to feeling so alien even to these 'user-friendly, technologies', we can become clumsy, panicky and silly very easily. Things that were supposed to ease life were actually taking lives due to user faults. All of these incidents – traffic accidents, exploding pressure cookers and water heaters – were actually intensifying the historical belief that the technoscape is a foreign territory, of which we want to be part, and yet it pushes us back to where we were (see Mungan, 2000). These accidents fed collective feelings of inadequacy and anger.

Sometimes, lived stories can tell more than any theoretical argument can express. Recounting the tragic story of the first national car highlights the historical beliefs, feelings and imaginations attached to the technoscape in Turkey. After the first coup d'etat took place in 1960, the new president, Cemal Gürsel, who had a military background, gave the order to produce a local car in a span of four months. Devrim (revolution) was to be the first national car and would be a symbol of Turkish independence, to show Turkish science and technology to the world. While engineers were working on the Devrim project at high speed, people ranging from academics to journalists to businessmen were publicly expressing their scepticism about the possibility of producing an engine or a car in Turkey (Şimşek cited in Bursalı, 2006). At the end of four months, four prototypes were manufactured and all were handmade. The outside look of Devrim was similar to an American car, which were generally purchased by wealthy families in Turkey at the time, but priced far cheaper (Bursalı, 2006).

On Republic Day in 1961, Cemal Gürsel climbed in one of these four prototypes to take Devrim on a tour to Anıtkabir (the mausoleum of M. Kemal Atatürk). He started the engine, but after 100 meters of driving, Devrim stopped. It simply ran out of petrol. It was not a technical problem, just that in the excitement and panic of public pageantry someone had forgotten to put gas in the car. In public discourse, however, the absurdity of the event had turned into futility (see Kortun's interview with Öğün – footnote 41). The 'failure' of Devrim was considered as something that shows our inadequacy in technological production. Now this story is often recalled as a 'failure' of Turkish technology. Once absurdity had become futility in the public eye,[6] Devrim was never manufactured again (Bursalı, 2006). The simple oversight engendered a much larger discussion in which the urge to express and explain the 'great lack'[7] of 'Turkish mentality' was collectively felt by so-called opinion leaders (journalists, columnists, etc.).

The impossibility of imagining a Turkish technoscape precluded the mass manufacture of Devrim, or any technological production. Turkey's love of technology largely became a practice of appropriation rather than production. Moreover, this love has always been accompanied by the body's exilic feeling in the technoscape. It was only after collaborating with European engineers and institutions that Turkish businessmen gained the confidence to produce another national car; Anadol was manufactured with the collaboration of the international automotive sector in the late 1960s. Technological production is intrinsic to the politics of the international economy. Economically marginalized countries are inevitably pushed to appropriate foreign technologies and hybridize them with local production of supplies. However, collective beliefs, embodied historical knowledge and emotions play crucial roles in the use of technology as well. The story of Devrim clearly illustrates the significant role emotionality plays in the social and cultural imaginings of the local technoscape.

The perceived strangeness of technology also contributed to understanding space and time as social and cultural constructs. Seyhan Erözçelik, another Turkish poet and a voice of collective belief and imagination, once wrote: 'how can a nation who is not capable of producing an elevator soar?' (cited in Mungan, 2000). The feeling of living on the outskirts of technology, residing in exilic zones of the international technoscape, has always prevailed when it comes to producing technologies. Consuming and using 'others' technology' have historically overlapped with a collective anger, a feeling of being duped, being a fool and a servant (Gürbilek, 2003) in the strange lands of technology. This is echoed in a number of journalistic representations which present the popularity of cell phone use in Turkey as a contested economic slavery to the western-dominated mobile technology industry.[8]

Although Turkey adopted many technologies almost simultaneously with Europe, the feeling of being late to the land of technology has never left its landscape. This sense of belatedness also draws upon desires, aspirations and fears, envies and resentments (Gürbilek, 2003).[9] Perhaps motivated by the fear and anxiety of being latecomers, and by a 'guilt' inherited from the lost past, any adoptable technology has entered people's lives at high speed. Dishwashers, video players, video cameras, and cassette players have all been integrated into the home, although most of the time they function as mere objects of decoration. I remember my grandmother used to have a cassette player and video player, which she never used, and a dishwasher, which she used only as an extra shelf.

In the 1970s and 1980s, music cassettes and cassette players along with video cameras and players were popular technologies in urban Turkey. The social structures of urban sites were changing due to internal migrations from rural regions, and some technologies and their related cultural industries were booming. Nurdan Gürbilek (2007) writes that in the late 1970s and early 1980s, crowds of people would contemplate shop-windows in which cassette players, microphones and any other available music devices were displayed. She suggests that these shop windows offered materials for collective fascination; 'technology' was something to be seen, contemplated and stared at (2007:33). Its actual use was limited to a small, wealthy segment of society.

Cassette players and music cassettes had become an integral part of domestic space on collective level, while Arabesque (the music genre) grew in popularity during the 1980s. Arabesque music is a synthesis of rural folk genres, western pop and belly-dancing music with strong oriental inflections. In his expansive study on Arabesque culture, Martin Strokes suggests that Arabesque is infused with the subversive sense of the periphery, with turbulent and violent emotions and with themes of alienation and powerlessness (2002: 214). The rise of Arabesque music, along with other cultural products like film and video, has been a significant part of the new public life in which historical melancholia has finally become consciously represented. The 'third world' (see chapter two where I speak about Gürbilek's use of the phrase 'third worlds' of Turkey) of Turkey was publicly raising its voice through technologies. As discussed in the previous chapter, the country adopted liberal-economiv policies in the aftermath of the last coup, and consumption has since become the new space for collective intoxication. The liberal market economy and the force of commodification paradoxically liberated the technoscape in Turkey. It becomes apparent that the 'third worlds' of Turkey is overlapped with the boom of cultural industries. The boom in cultural industries and the expanding production of multiple cultural products also amplifies the voice that has learned how to speak in order to speak. That is, the unspeakable issues – historical losses, the violence of the state, the wounds and markings of the last coup – continue to delimit what can be uttered publicly in cultural products like films, Arabesque music, radio and television programs.

The motto of the 1980s, 'Turkey is opening to the world', implies the role given to technologies in this ambition. Turkey was becoming another country, shifting from an inward- to an outward-looking social structure. The organization of a national technoscape was seen as crucial for achieving the dream of making Turkey a body that could communicate and be compatible with its western counterparts. 'Highways are liberty', announced Turgut Özal, the Prime Minister at the time, who initiated the liberal market economy and started the privatization of state institutions.[10] He promised new highways, a high penetration rate of telephone lines and electricity, and water to areas that have always been economically and socially marginalized (the Kurdish districts especially). Transforming the technoscape was a necessary part of imagining and idealizing a civilized, modernized and unified Turkey.

Where the state was unable to unify, mobilize and modernize the country, commodification would do the job. Commodified spaces and practices were there to externalize the collective losses and disappointments, to open up sites where bodies could do things, could move in opposition to the cultural policies of the state. The commodified technoscape started to run counter to the state technoscape. As radio was the voice of the state's strict modernizing policies, commodified sound technologies enabled alternative social practices of listening (see Özbek, 2002; Markoff, 1994).[11] While state-run television was a medium of pedagogical state-power, video technologies provided entertainment. While telephone technology was restricted to the privileged, wireless emerged to integrate the scattered agents of the periphery. In opposition to the state-run technoscape, the commodified technoscape was pointing to the possibility of intimate experiences between the social body and technology.

Imported technologies and recording techniques also created variations in popular music. Orhan Gencebay, for instance, used the electronic *bağlama* (the traditional guitar of Turkey) and other technologies in his music, which was most popular in the 1970s and 1980s (Özbek, 2002). Music listening technology became more desirable to the general population with the addition of Arabesque music to the commodified technoscape. Finally, 'technology' was speaking a familiar, understandable language to which peripheral masses could relate. The collective feeling of technology as fascination and anxiety was now accompanied with pleasure and entertainment. Music cassettes and cassette players offered a different mode of engagement that partly broke the cultural distance between sound technology and the social body that was constituted by radio broadcasting. People were free to listen to whatever they wished outside of state-run radio. While the radio music programming was only offering a repertoire of Turkish pop music (an adaptation of western pop music), classical Turkish music (historically called palace music), Turkish folk music and western music, music cassettes and cassette players generated experiences of the technoscape that were not ruled and programmed by the state. Arabesque was banned in radio broadcasting because it was viewed as backward and Arabic-oriented for a long time (see Meral Özbek's 1997 work on Arabesque culture).

In a similar manner, the wired telephone system belonged to and was solely administered by the state. From the 1960s through the 1980s, the people of Turkey had to wait for years to have a telephone line (see Geray, 2002; Törenli, 2005). Officially, people with certain occupations – such as doctors, journalists, politicians – had priority, yet everyone knew that these rules were only applied to those without connections in state institutions. The telephone system also lacked a physical infrastructure. Some residential districts in large cities were excluded from the wired system, as were some regions of the national territory – like the south and east regions of Turkey. The strict dissemination of telephone services intensified feelings of resentment, anger and envy on a collective level. While telephone technology could only be obtained by the privileged, wireless technology has become a distinct social practice that integrates scattered agents into a group of 'friends of wireless'.[12] After the long ban on wireless, the government passed a law in 1983 that permitted its use by the public. By paying a very small amount of money for the permit, people were able to communicate with each other through wireless. The expression 'looking for a friend, for a friend'[13] came to define wireless technology in Turkey in the 1980s. Sconce writes, 'as wireless put more people into contact, it did so with a sense of melancholy' (2000: 63). For Sconce, wireless was not only a form of contact for scattered people around the world, but also a reminder of one's loneliness and isolation. Sensing social and cultural segregation and separation, the limits of which were delineated by the state, was crucial in constituting this 'melancholy of wireless.'

After a long period of state-run television and radio (TRT), a law allowing private radio and television channels to broadcast was passed in the early 1990s. Throughout the 1990s, new radio channels were introduced almost every month and new television channels entered the mediascape almost every year. 'People found their tongue' largely through radio and television broadcasting (see the previous chapter). Not only were these new media opening up possibilities for deciphering the private lives of people, but they have also become significant political tools for Islamism. Discussing how Islamist politics have become so powerful in Turkey in the last two decades ¢erif Mardin emphasizes the role of technologies, particularly television and radio. He argues that Islamists have always been keen to adopt technologies for their own pragmatic ends. Using mass media as the backbone of a political network has always been a central to political Islam in Turkey (Mardin, 2005). Diversity in the mediascape, however, has generated and intensified the fragmentation of society into different political camps. The mediascape has been divided between Islamists' and secularists' television and radio stations just as print media has been. Now in the 2000s, ultranationalism has become another popular voice across the media.[14] All of these examples of the rise and use of technologies demonstrate how crucial technologies are to social and political struggles in Turkish social space, where the social is lost to the enclaves of various social and political groups. Yet the technoscape still holds promise for unity and social bonds as well as for articulating individual identity.

These commodified technologies shaped sites where the marginalized can speak as well as hear themselves. Not only did they open up a possibility for migrants, the poor, children, youth, or any 'other' in society, but they also engendered new social practices in which urbanized, middle-class people found ways to express and recognize their own 'third worlds' (the other in the self that has long been suppressed by the strict modernization policies of the state). Video technologies, music listening technologies, and television, among other technologies, concomitantly produced distinct social practices for different individuals that familiarized 'technologies' with 'culture', with what is assumed as our own. These technologies were imagined and inserted in social practices of everyday life which helped to make people's 'third worlds' recognizable, acceptable and even loveable both by themselves and by others. By using the power and authority of technology, 'third worlds' could become more decipherable for the social body. Appropriating technologies for the production of 'local' has worked to make 'local' or 'what is our own' acceptable to show.

Recognition of the other did not occur peacefully in social space. For instance, while 'friends of wireless' became a collective through their exclusion from the state-run telecommunications technoscape, wireless also came to define a certain class' taste and habitus which was degraded and considered backward by elitists. Technological practices have always been integral to social struggle. More importantly, social practices of technologies in Turkey trigger the historically embodied desire to 'express' or to 'show' what we, as people of Turkey, are lacking. The technological habitus of the 'people of Turkey' has come to reflect and reproduce a Turkey imagined as in between the east and the west, not as the desired 'bridge' but rather a disgraced image of 'homelessness', the landscape where a collective inferiority complex prevails.

Journalistic representations of cellular telephony continue to reproduce a derogatory attitude towards the so-called periphery's habitus. One newspaper column states that technology functions as a mirror in which we can see ourselves objectively, and our people's use of cellular telephony shows what we lack compared to any western country; in essence we are ignorant (Berberoğlu, 1999). According to this common understanding, not only do we lack the scientific reason to produce technology, but we also lack the very knowledge to use technologies properly. Yet paradoxically the technoscape remains the force that will transform lived Turkey into the landscape of imagined Turkey.

The technoscape: an historical dream of presence

In this work, I approach the concept of technoscape by borrowing Rose's conceptualization of the landscape as a dream of presence. Rose (2006) suggests, 'the utility of thinking in terms of dream of presence [is] that it identifies those named closures [that] look like culture…as indicative of specific horizons of being that are perpetually moved towards but never reached' (545). In this sense, the technoscape is not a kind of 'arrival', a closure that looks like culture, rather it is a process of imagining that engenders becoming. My contention is that the technoscape is also 'real' in that it binds 'various inclinations, sensations, and responses into particular imaginations of presence' (Rose, 2006: 545). For Turkey, the technoscape as a dream of presence imbricates a connection between self and other, body and collective, body and world, and local and global. The technoscape is a 'performative and emotive' (see Reddy, 2001) dream of presence which provides infrastructures, communication pathways, narratives and stories to orient and inspire becoming for a collective as well as for an individual. As a dream of presence, the technoscape allows subjectivities and culture to be experienced. The failure of the technoscape in its arrival to 'subject', 'culture', 'unity' and 'home' is integral to the performance of the technoscape.

The performance of the technoscape can be thought of as realizing what Derrida (1987) terms the 'postal principle'. In the 'postal principle' the movement is one of arrival to the imagined and impossible 'self-presence', 'home', 'the proper'. An imagined presence conditions the movement. As Derrida maintains, because the arrival is impossible (because the ideas of subjectivity and culture refer to impossibilities), the play, the movement, the travel or the itinerary is possible. The whole itinerary appears in Derrida's *The Postcard* as 'an effort to love oneself', to find oneself in the proper, at home (1987: 21, my emphasis). Movement is an effort to get back what has been taken from the being. Everything becomes a postcard for this movement, Derrida says. The postcard is what precludes the arrival; that is to say the 'courier' (any courier: technology, language) is a barrier to 'the communication process, or an obstacle for destination' (ibid.). The arrival is impossible, however, its possibility in imagination provokes both movement and its iterability or repetition. As a dream of presence, arrival calls on us to 'depart' from, to 'exit' from, to 'erase', to 'alter' what is at hand, like the self, the landscape, the culture. The presence of the courier in forms like technology, a tool, or a language, shows that the arrival has not yet occurred, that something is still missing. For example, Avital Ronnel suggests that the telephone marks an absence, that it 'fundamentally belongs to the sense of not-being-at-home' (1989: 45). The telephone, the language, the postcard, the technoscape are not given up because they 'realize' the arrival in the imaginary, they 'accommodate the lack' in the fantasy (Wills, 1995).

As an historical dream of presence, the technoscape has performed its task in Turkey, of accommodating the landscape's presumed 'lack'.

The gap between the Ottoman and European technoscapes 'marked' the time-lag between the east and west, for the technoscape has historically been a site that 'shows' temporal differences between the two. As such, it has also been a site of valuable investments, economically and affectively. The technoscape has always been both the source and the means of overcoming melancholia. It has been a site where the 'lived' is intolerably recognized and the 'non-lived' (the dream) continues to be imagined (see Hansen, 2004). Not only has the technoscape been crucial in terms of 'showing' the irrecoverable distance between the east and the west, now the local and the global, but it has also played a key role in Turkey in demarcating state and society (see above), elitists and the 'people', Islamists and secularists. In short, the technoscape has always been crucial to the political, the cultural and the social in Turkey. As Ronnel (1989) suggests, technology has ruled – and continues to rule (81).

The technoscape is where different agents feel and imagine together. It is also where differentiations and distinctions are contingently organized. The technoscape is both a material and an imaginative space where historically embodied knowledge and belief – the habitus – comes through in every practice. It is also integral to the organization of social space, where the 'taste' and the types of 'capital' determine the very position that an agent assumes. The technoscape in Turkey is necessarily both bound by historical imaginations and open to the potentialities of new experiences, asserting through bodily sensation the possibility of 'change', 'movement', and 'attachment'. There is no one national technological habitus in Turkey, though embedded idealizations and feelings have become a second nature that affects bodily technological practices.

In this work I think of the cell phone as a particular performance of the technoscape. The concept of technoscape as an historical dream of presence provides a constructive space for thinking of cellular phone performance in relation to past imaginations of the technoscape, and to other contemporary performances as well. That is to say, outside of imaginations of the technoscape (as what is 'believed' to transform the landscape), and outside negative imaginations of the technoscape (as what is presumed to show the 'time-lag', the 'backwardness' of the landscape), cellular phone performance cannot be effectively and productively thought. Considering the technoscape as an historical and continuous dream of presence enables me to situate cellular telephony as an integral part of the technoscape – as a technology which is both conditioned by memories, social structures, beliefs, knowledge structures and collective feelings – and which mediates the bodily experiences of the technoscape. I conceive of the relation between cellular telephony and the technoscape as continuous, within which historical imaginations and idealizations affect practices and are also subject to their own effects. My work attempts to comprehend how cellular telephony is situated and implicated in social struggles, how it corresponds to the past experiences of other commodified technologies in Turkey, how it speaks to the 'third worlds' in us, and how it engenders a collective addiction. This framework elucidates the conceptual and practical links through which cellular telephony can be grasped as a social practice, replete with meanings, functions, emotions, and affective and imaginative potentialities.

4

ATTACHMENT TO CELLULAR TELEPHONY: THINKING OF MEANING, FUNCTION AND BODILY RELATIONS

> If I don't have my cellular with me, I feel empty. I can lose my lover, but I don't want to lose my cell. My cellular is my everything![1]

> Even if I don't use it, or do not intend to use it, I want it to be near to me...actually I want it to be in my hand. (Meryem)[2]

> I think if there is a nation that would never detach from cell phones, it is Turks. Perhaps we should change our motto from 'horse, woman, weapon' to 'horse, woman, weapon and cell phone'. (M. Ali Yilmaz, Milliyet, 2004)

Answering the question of why and how the cell phone has become an object of collective attachment in Turkey entails a description of the nature of this attachment. This chapter specifically aims to deal with attachment by presenting some firsthand accounts of people's attachment to this artefact in Turkey, and proposing certain interpretations of their observations. After quickly reviewing how current literature on cellular telephony largely considers the technology in terms of its meanings and its instrumental work, I question to what extent I can claim that the cell phone becomes an object of collective attachment because of its meanings or symbolic value, and its function or instrumental value. Moreover I comment on its commodity aspect and examine to what extent I can consider the collective attachment to cellular telephony from the standpoint of an addiction to consumption. I conclude the chapter by arguing that the current attachment to cellular telephony in Turkey urges us to think of the cell phone as an artefact that produces affective, psychic and social effects, and that these have become linked to the heart of life, as if the cell phone generates a life-giving activity for people. To conclude this discussion of the bodily attachment to cellular telephony, I comment on the ways in which the cell phone is incorporated by the body through hands (or the work of the hands).

Defining a collective's engagement with cellular telephony as both an attachment and an addiction describes how, as a technological tool, the cell phone has been transformed into an object of intense relations and compulsive use. At present, addiction is generally seen as the habitual or repetitive use of an object, and seems to have lost its conceptual link to the notion of attachment in popular discourses. Addiction traditionally denotes an intense attachment and devotion to an object.[3] From this perspective, attachment leads to habituation, which may end in addiction; this is tantamount to saying that addiction is a compulsive attachment. When understood in conjunction with attachment, addiction designates a body's intense attachment to an object without which the body feels it is missing something essential, and accordingly shows withdrawal symptoms such as anxiety, discomfort or even depression, an emptiness in the object's absence. My following analyses aim to show that the collective attachment to cellular telephony in Turkey should be understood as an affective – though not necessarily emotional – attachment in which bodies invest in order to tell specific, special narratives to both themselves and others.

Instrumentality, meaning and attachment

Current studies consider attachment to cellular telephony in two main ways. First, researchers claim that as a technology the cell phone generates instrumental value for interpersonal relations (Ling and Yttri, 2002; 1999; Katz, 2002; Fortunati, 2005; Ito, 2005; Okabe et al. 2005). Secondly, studies show that the cell phone acquires meaning as it represents one's social status and particular lifestyle (Vershinkaya,

2002; Chakraborty, 2004; Katz and Aakhus, 2002, Ling, 2003; Donner, 2005a; 2005b; Özcan and Koçak, 2003). The cell phone's instrumentality is largely discussed in terms of its structural possibilities – the kinds of work it can do for interpersonal relations. As I have shown in the first chapter, researchers have concluded that the cell phone is most often used to increase personal security, and to facilitate coordination and social interaction with others. The cell phone is basically considered a telephone that can be carried anywhere the body goes, making the body available anytime and almost anywhere. Its most instrumental function is point-to-point communication, although it can also be used as a broadcasting medium.

Apart from telephonic interaction, research shows that short messaging services are highly instrumental and useful for people who prefer cheaper communication services and/or are more comfortable writing rather than speaking on the phone (Elwood-Clayton, 2005; Pertierra, 2005; Strom, 2002; Taylor and Harper, 2003). Other research illustrates how texting can work as a mass medium to gather a crowd for public demonstrations against the state or any other authority (Hill, 2003; Lallana, 2004; Elwood-Clayton, 2005; Sheller, 2004; Katz, 2002; Plant, 2000). In addition to texting and telephonic interaction, cell phones are used for a variety of entertainment purposes such as recording life, broadcasting what is recorded, and sharing music/video/pictures with others (Rivière, 2005; Chesher, 2005; Richardson, 2007; Goggin, 2006). The cell phone is proving to be highly instrumental to everyday life, especially in relation to elements of security, coordination and social interaction.

The cell phone's social meanings also contribute to its popularity in many different countries. It is said that the cell phone as a personal object is considered by its users as a means to express one's identity both to oneself and to others. Its meaning in social life is largely related to how it demarcates social status and lifestyle. Accordingly, authors have dwelled on the aesthetic value of the cell phone, as a cool and fashionable object symbolically representing a desirable modern and mobile lifestyle (see the articles in Katz and Aakhus 2002).

In the first chapter I suggested that this interpretation alone does not furnish an understanding of why and how the cell phone becomes an object of collective attachment in any particular context.

Here in this chapter, I will argue more fully why attributed meanings or instrumental value alone cannot make the cell phone an object of attachment. However this does not mean that meanings and instrumentality are negligible. On the contrary, the meanings that the cell phone acquires in social space are crucial, playing a significant role in making the artefact appealing and attractive for possession and use. Likewise, the instrumental value of any technological object is highly important if it is to become ubiquitous, because if a tool does not provide aspects of use that can be easily adapted to everyday life practices, it would not be used by or attached to a body at all. Furthermore a tool needs to be user-friendly in the sense that its use does not require a new bodily skill or qualification. In short, a technological tool needs to be meaningful and useful, otherwise there would be no possibility for it to 'become' an object of attachment (Verbeek, 2005).

Yet meanings and instrumentality are only conditions for an attachment, which is to say an object with a certain meaning and instrumental value may not generate a bodily attachment. As we start thinking about technologies, we begin considering an articulation[4] which involves the human and non-human parts. The question of why and how the cell phone becomes an object of collective addiction necessarily requires us to think of the cell phone's functioning (its own weight)[5] and also its user's purposes, wishes, desires, imaginations, and feelings. As Peter-Paul Verbeek (2005), a post-phenomenological thinker, writes, 'the relation between human beings and artefacts comprises more than visible aspects…an artefact can play more roles in human life than functional ones' (30).

The meanings of cellular telephony: social status, lifestyle and beyond

As a marker of social status, identity and lifestyle, cellular telephony has traditionally been thought of as a defining habitus of a certain class or segment of society, and this attribution is reflected in and perpetuated by journalistic representations and scholarly works in Turkey. It has been suggested that the 'people', or the 'third worlds'[6] of Turkey use the cell phone to associate themselves with a modern-urban lifestyle, economic-cultural-social capital and the distinctive taste that would make them look like the upper classes, or more precisely, like ruling elites.[7] To support my contention that the cell phone's symbolic meanings are more inclusive than simply representing desired or actual social status, I will provide excerpts of a detailed interview that I conducted with two owners of a cell phone shop in one of the poorer districts of Ankara. This interview, along with my observations and interpretations, will show that the cell phone is not only a marker of desired social status for the 'third worlds' of Turkey but also includes other forms of cultural capital like 'agency' and 'individuality'. I will also suggest that attributing meanings to the cell phone in order to use the tool as a 'reflector' of oneself or one's desired self is not characteristic of a certain class but rather applies to all bodies that are attached to cellular telephony.

The interview I transcribe here was conducted with co-owners of a cell phone shop located on one of the main streets of Akdere. Akdere is a significant space in my study: it is populated by the so-called periphery, whose habitus of cellular telephony is said to be characterized by displaying the cell phone as a marker of ideal and idealized social status and lifestyle. This district used to be one of the main *gecekondu*[8] regions of Ankara, but every year it becomes more modern in appearance with new apartment buildings (though still not too many in the whole district), supermarkets which sell cheaper products, a variety of small shops, asphalted roads – at least in main streets – and traffic lights. Akdere has become more ethnically diverse over the last decade, due to the forced migration from the southeast part of the country. It also has a more right-wing political culture; Islamist and ultra-nationalist tendencies dominate political discourse. Most residents in this district, especially in *gecekondus*, subsist on municipal financial aid. There are ten cell phone shops in this district; all opened within the last five years and are still operational. A cell phone store in big cities generally sells cell phone machines, accessories, and pre-paid calling cards for various GSM operators. Some of these shops also repair broken or obsolete cell phones. The shop I visited sells brand new and secondhand cell phones and a variety of accessories for different models, and also offers a repair service.

The shop's owners, whom I will call Kamil and Cemil,[9] were raised in Akdere and began their business – which is across the street from their home – in 2004. Kamil, Cemil and I talked about a range of issues pertaining to cellular telephony in Turkey, such as unemployment and cellular telephony as a job market; social inequalities, class differences and possessing a cell phone; crime, robbery, the state and the second-hand cell phone market; technology and Turkey; and consumption patterns in Akdere. Because we spoke in their shop during a workday, our talk was interrupted many times by customers and visiting friends. As a result, our discussion also included my informants' assumptions about and interpretations of these people's tendencies with regards to cellular telephony, and observations about their consumption habits.

Kamil was unemployed before he opened this shop, holding a series of temporary jobs which did not require specific qualifications. He had his first cell phone during his military service, when he was a personal and private driver for a coroner.[10] He bought this machine, he says, mainly because he thought it was cool, and also so he could communicate with his family.[11] Although normally one is not allowed to have or use a cell phone in the military quarters, Kamil was free to use his because the coroner he worked for needed to contact him easily. Kamil says this made him more respectable in the eyes of the other men serving in the military and not possessing a cellular. During what was a depressing year of service, this toy of his had given him a sort of power/authority over other men, a chance to get rid of the feeling of isolation, and a sense of play as he spent his free time discovering how the machine worked. He told me that during his military service he became very familiar with the machine's technical and mechanical aspects. After he left the military, he started hanging out at friends' repair shops learning how to fix broken or obsolete machines. In the end he became a repairman himself and opened his own store by taking out bank loans with his partner. Their shop has been robbed twice, and each time approximately 50 machines were stolen. Neither the thieves nor any of the stolen machines have been found. Kamil says they have a difficult time feeding their families yet they are still happy to at least have a decent job.

As Kamil's experience shows, cellular telephony in Turkey has created employment opportunities for many who did not hold permanent jobs before. The second-hand market has also grown large enough that people can earn money by repairing and reselling cell phones, although this market is largely comprised of stolen cell phones.[12] Not only does cellular telephony provide jobs for the unemployed (both legal and illegal ways of making money) but it has also become a big retail sector, shoring up the country's economy to a certain extent. The growth and size of the cellular telephony market also means that a 'technology' can be fully domesticated in this country.[13] Columnist Mehmet Ali Yilmaz's observation at the beginning of this chapter expresses the common perception that cellular telephony in Turkey has become a defining symbol of national attachment. Chapter six will deal with this issue in greater detail; for now I will address what Kamil has said about the 'meanings' of cellular telephony.

Kamil told me that in the first years of cellular telephony in Turkey, he renewed his cellular continuously and loved to have fancy models. He said it was because:

> When you put your cellular on the table[14]…they say 'Wow, see Mister Kamil's telephone.' When they do this…you get proud… because they don't expect Mister Kamil to have something like that…you know, we don't mean to look down on other people… our intention is not to show off…but this machine shows what kind of person you are…

When asked if this has anything to do with 'showing' his financial success, he replied:

> No…I'm not rich and you don't have to be a fortune-teller to guess that I'm not…The cell phone is not about showing money…but it is like, yes we are *kıro*, but we know how to use this technology, we know which model is good, which is the best…the issue is not whether you have money but what you do with your money.

Kamil says, 'yes', he is a *kıro*, meaning that while he may not be considered a cultured and educated man according to current social categories in Turkey he knows how to use this technology, he has a certain taste, and is capable of discerning which model is better than another. He displays his cell phone with the intention to 'show' what is not expected from him, 'because they don't expect Mister Kamil to have something like that'. In Turkey's technoscape, the periphery's technological habitus has historically been seen in one of two (elitist) ways. Either 'people' (in the strict sense of 'third worlds' of Turkey) are totally absent from the technoscape, for they lack the economic and cultural means of possessing and using these technologies, or if they are familiar with particular technologies then they do not know how to properly integrate technology in their everyday lives (see the previous chapter).

Kamil has clearly encountered these hostile 'expectations', the hurtful prejudice that does not derive from a discrimination based on poverty, but rather the assumption that one does not have a claim to 'agency',[15] is not capable of adapting to the requirements of contemporary life, is ignorant, lacks necessary life skills. In chapter two, I showed how important catching up with the contemporary world is for the historical imaginations of an ideal Turkishness. Likewise, in the previous chapter I outlined how technologies are historically imagined as providing what is assumed to be lacking or missing. Kamil's statements can be interpreted in the light of my previous explanation of how social structure, historical beliefs and collective emotions are embedded in one's body and expressed in one's actions. Kamil voices the common prejudice towards the 'Third World' and affirms his interpellation as part of the periphery by saying, 'yes, I'm *kiro*'. Yet he also says, 'but I know how to use this technology, which model is better…' Kamil's performance of cellular telephony explicitly confronts the discriminatory 'expectations' that are embedded in the social structure. Cellular telephony thus does not simply represent social status, but also articulates a felt sense of 'agency' in and to himself and others. Or more precisely, cellular telephony transforms its holders from being 'objects' of elitist negative feelings (see Ahmed, 2004; see also chapter three), to agents capable of asserting and claiming their share from the technoscape.

The elitist vision of the technoscape posits Islamists as having lifestyles and/or political, cultural and social tendencies that are irreconcilable with the goals of modernity and contemporary ways of living. Yet urban Islamists struggle to show that there is no singular way of understanding modernity nor is there simply one way of adopting modern lifestyles as a public part of their identity politics. In an effort to explain that

Islamists are also modern agents of this landscape, Ali Bulaç, a prominent Islamist intellectual who has theorized extensively about modernity, writes that urban Islamists were the first in Turkey to incorporate new technologies like the cell phone into their everyday lives (1998). He argues against categorizing the conservative Muslims[16] of Turkey as the ones who reject modern life practices, suggesting instead that they are as eager as secularists to embrace such ways of living insofar as these practices promise to bring new, meaningful and functional experiences to life. Like Kamil, Bulaç sees cellular telephony as part of collective as well as individual struggles in social space. These understandings of cellular telephony confront the discriminatory beliefs and knowledge about Turkey's 'third worlds', and illustrate how this technology comes to show that such assumptions are not necessarily true.

When searching for stories of cellular telephony in newspapers, I came across an interview with Emrah, a well-known Arabesque singer who had become incredibly popular as a little boy in the late 1980s. He describes how he has changed since then, both as a musician and as an individual. He invokes cellular telephony as a positive symbol of change, saying:

> I had to change and I have changed. Otherwise it would be like saying, 'the cell phone has been introduced, but I don't use it'. They will beat up the ones who do not catch up with the requirements of today's world...One's personality changes. If one doesn't change, then there is a problem. I have changed as well. But the issue is changing in a good way. This is the secret of life success. (*Hürriyet*, 1998)

This analogy between embracing cellular telephony and personal change is significant, as it points to the very 'invisible' relation between technological engagement and self-articulation. Similarly, renowned journalist and documentary film-maker Can Dündar, in his essay on the stereotype of Turkish *delikanlı* (which in some cases is used interchangeably with *kıro*) writes: 'The formulation of "a rosary in one hand, a cellular in the other" has found great resonance among the conservative youth and middle classes in the late 90s. In the guise of merging the traditional with modern, there lies an aggressive, hostile concealment of the wound...' (*Milliyet*, 2004). I do not agree with Dündar's interpretation of a rosary and a cell phone together in *delikanlı*'s hands as a 'concealment' of the 'wound'. Rather than understanding *delikanlı*'s attachment to cellular telephony (or that of 'third worlds' in a broader sense) as a form of aggressive concealment, I would contend that it can be more compellingly read as an aggressive and insistent search for a 'new' presentation and articulation of self that can gain social recognition from others.

These sorts of statements and suggestions emphasize how the cell phone becomes an expressive artefact – part of the narrative that an individual or a collective wants to tell about herself/themselves, both to oneself and to others. As Zoe Sofia, a philosopher of technology, writes, 'the tool is never a purely material object, but always has its partial origin in the inner world, which is to say that it is always meaningful, part of a narrative or set of human purposes' (2000: 186). The purposes may vary in character, yet cellular telephony clearly comes to function as a part of narrative that is told to oneself and to others simultaneously. Just as presenting and articulating oneself via cellular telephony can challenge the discriminatory 'expectations' of others, 'moving' oneself away from being an object of negative feelings, so too can it function to differentiate oneself in social space. Its work can be interpreted as a performance as well, one that becomes meaningful insofar as it promises a certain sameness or commonality between different collectives. Kamil, Emrah, and Ali Bulaç's comments convey how an engagement with cellular telephony is also a social practice through which these diverse, polarized groups can form and display certain social bonds. Vicente Rafael (2003) invites us to think of cell phone politics as a messianic discourse replete with promises of justice, commonalities, an integrated collective and a crowd of equal people; I explore this sense of the social promise at stake in cellular telephony more thoroughly in chapter six.

Ascribing to cellular telephony the ability to 'reflect' oneself – or more precisely one's desired self – is not a tendency specific to Turkey's *kıros* or 'third worlds'. This habitus does not define a certain class' relation with the cell phone; rather it applies to all bodies attached to cellular telephony. If technological habitus defines people's practices with an artefact, then the very habitus that we are talking about contributes to an articulation of self, body, and individual that demarcates one's difference from others in social space. One of my interviewees told me that if she needs a new cellular, she goes to the market and gets the least attractive one. She says, 'because I hate the fact that it is over-valued in a way that people continuously want to buy the most fashionable one. I'm not one of them and my cellular shows that' (Emine).[17] The cell phone acquires a meaning relating to and becomes part of the narration of self. As another woman tells me about her cellular:

> It contains all of my memories, my photos, messages, telephone numbers of my friends…it is like me one can tell what kind of person I am by looking at inside of my cell… (Eda)[18]

Cellular telephony offers more than a display of one's desired social status and lifestyle; it articulates and reflects these visions of self and individual identity by promising a certain sociality as well. Social status, lifestyle and economic, social and cultural capital are all integral to individual identity and to cellular telephony's meaning for the self. Despite rigid budgets, many people in Turkey purchase this object and renew it whenever they can, making both economic and affective investments. Understanding the extent of users' desire for cell phones might help us to grasp the intensity of this attachment. To show the most striking and illuminating aspects of people's investment in this technology we need to look closely at the tendencies of 'third worlds' to consume this product. Once again we return to Kamil's shop.

In the course of my talk with Kamil and Cemil, two of their friends came in and left a machine there for repairs. These friends hesitated to say what happened to the machine; apparently my presence created a sort of discomfort. They eventually confessed that the machine fell into the toilet. After they left, Kamil turned to me and said:[19]

> See these people that have just gone…they are my friends…one is a driver, the other one is a guard guy in a private company… the telephones they have cost more than they make in a month… but thank to credit cards, they buy it…and after that they can't pay it, then they take bank [sic] loans to pay the credit card fee… it goes like that.

Kamil tells me that people unable to afford food or cover other vital expenses will still purchase a cellular and fill the machine with pre-paid credits.[20] He says:

> It is unbelievable but true…you know, this is a country of poverty, unemployment…See, in the last couple of years violence, robbery, crime have incredibly increased because of poverty, inequality, imbalance between different segments…but the cell phone is something that never loses its value…No matter what it costs, no matter what expense to them, people buy it and continue to renew it.

In an insightful work on the social resonance of the cell phone in Turkey, Asuman Suner (2001), a media and film scholar, searched for an answer to the question: why, in a country where the income per capita is very low compared to other OECD countries has the cell phone become as popular as it is in some European countries? Rather than look to more traditional formulations of social status or modern lifestyle, she suggests that it is perhaps because the cell phone gives its owners a feeling of movement, departure, or migration to another time and space. She writes that in a landscape where the majority of young people dream of leaving Turkey for somewhere else, the cell phone's associations of mobility, liberty, and freedom speak to the imagination and virtual experience of departure.

In the beginning of the 2000s, the market-leading GSM provider Turkcell started an ad campaign that has become both popular and the subject of controversial debates in the popular media. In this series of ads, a young woman, with an international traveller look[21] wanders around 'exotic' places in Turkey, enjoying her freedom and liberty. The concept clearly linked freedom to cellular telephony. These ads specifically promote Turkcell's pre-paid calling service as a sort of 'freedom', implying that one does not have a responsibility to pay the bills – once you get credits, you are free to talk. After this campaign was critiqued by intellectuals[22] for depicting the southeast part of Turkey as an 'exotic' or an 'other' place and for its 'Orientalist tone', a director of Turkcell's advertising agency offered a number of justifications for this approach. He argued in part that the agency was working from the belief that many people in Turkey have a strong desire for freedom, particularly as it connotes notions of liberty, mobility and travel.[23] In this case freedom even implies a sense that one can see the 'problematic' landscape with the eye of a westerner who has no responsibility for those suffering there, no ties to them except as a tourist. Regardless of what these ads actually mean to their audience, they invoke a powerful promise of cellular telephony.

Ronell (1989; 1995), following Derrida, argues that technology promises a certain exit and exteriority, a departure from the self at hand and the place where one is to someone and somewhere else. She writes that in this way technology likens itself to a drug, in that both of them 'generate a kind of a hallucinated exteriority, with them the distinction between interiority and exteriority is radically suspended' (1995: 65). Both of them are highly addictive. Kamil cogently articulates the core of this issue:

> Look, this is the 'king' of all technologies in our time...not only does it show what kind of person you are, how important you are, but it is also useful...you can do almost anything with this machine...Look, people first buy it because they think that it is cool, but then it becomes a sort of disease, a mania...you become addicted to it...Take for instance the guys that came here, who dropped a cellular in the toilet...people carry it everywhere even to the bathroom...it is not only about impressing other people... Yes, looking cool is very important, but, I mean this guy did not want to impress anybody in the bathroom, I guess, right? So it is more than charisma.

Just before he said this, we had talked about whether or not the cell phone's basic promise is to endow its users with 'charisma'. He did not specifically use the word 'charisma', which we also use in Turkish; he called it 'şekil,' a slang term referring to how hipster or cool or impressive one looks. I am not suggesting that cellular telephony has no display value, nor am I discounting its relation to how aesthetic values come to reflect one's taste or desire to look more modern, mobile, wealthy or cultured. As I argued above, these meanings have contributed significantly to making cellular telephony attractive. When the cell phone was first introduced in Turkey in 1994, it was defined as a personal technology. As I have discussed in the previous two chapters, technology is a sort of messianic word for melancholic bodies of Turkey. Therefore, it was implausible that the people of Turkey would be indifferent to a personal, purchasable technology. Yet these reasons only explain why and how many people of Turkey opened to their bodies to this technology, they do not tell us about the nature of the attachment.

Verbeek (2005) writes, 'objects can acquire their meanings for people first of all from their intrinsic aesthetic qualities. That is true, naturally, for artworks, but also for useful projects in general to the extent that people find their exteriors pleasing or that they express a particular lifestyle with which people want to be associated' (223). According to Verbeek, the body first opens to an artefact through an appreciation of its aesthetic qualities and its socially constructed meanings. This allows users to be open to the product, but in principle other products might also serve the same purpose. 'Besides,' he notes, 'lifestyles and the signs that express them are heavily influenced by fashion. What things do or mean for people is a role that can also be filled by other things – with the exception being those things to which people are strongly attached' (232). Such exceptions are built upon attachments that arise out of the memories and associations that develop over years of use.

Today, the majority of Turkey's population of Turkey owns a cell phone (see chapter one) and many express that without one they feel lacking in something, as if they miss something essential to their bodies. After identifying the cell phone's function in narratives of the self, its propensity to symbolize a particular lifestyle, its role in social struggle and social space, and its representations of freedom and liberty, how do we talk about the cell phone as a commodity? Can we suggest that it sells the feeling of travel, freedom, prosperity, individuality or particular lifestyles? Or, more precisely, can we say that the addiction to cellular telephony is an addiction to consumption of what it means and what it does for us?

Cellular telephony as a commodity

Albert Borgmann's (1984, 2000, 2006) analyses, which consider contemporary technologies (he calls them devices) as first and foremost commodities, are helpful for thinking of the attachment to the cell phone as an attachment to a commodity. For Borgmann, technological devices consist of two elements: machinery (the device as a physical object), and the commodity the machinery delivers when functioning. For instance, a central heating system is a device. Its tubes, radiators and heater are machinery and heat or warmth is its commodity. Borgmann writes that 'what makes something a commodity' is its 'structure and construction', which have social foundations, rather than its 'interpretation' by its users like its availability for 'self-projection' (1984: 77). If we think of the cell phone as a device, the base stations, keyboards, screens, plugs are its machinery and communication is its commodity. Borgmann notes that insofar as a device is useful or user-friendly it becomes a commodity, while its meanings and interpretations make it a highly commodified device (1984, 2000). Being a useful, user-friendly device means that the machinery becomes invisible to the consumer, it leaves her/him outside of its functioning and simply provokes consumption. In other words, devices do not have the capacity to engage. One consumes only the commodity that the device provides, and is neither involved in its machinery nor able to 'appropriate the device through care, repair, the exercise of skill and bodily engagement' (1984: 48). Borgmann argues that devices disengage human beings from their machinery, and accordingly from material culture and even from reality. Since a disengaged user cannot participate in the device's function, renewal, or refashioning, s/he is deprived of the pleasure of engagement and is left with the fickle pleasure of consumption that is free of preparation, exertion and obligation. He writes, 'addiction is the most severe case of the evaporation of pleasures...the pleasures of commodities fail to keep their promise. They lure us to ever renewed efforts, but they leave us unfulfilled and betrayed' (2006: 356). This addiction is, more precisely, an addiction to consumption, for since devices are mainly commodities, an addiction to them is an addiction to consumption.

Borgmann's analysis emphasizes that we need to understand how the cell phone functions as a material object in relation to the human body – whether or not it allows its users to be involved in its machinery, and if its machinery becomes completely invisible. As a device, the cell phone can be used easily, and does not require specific skills beyond knowing how to make and answer calls. The cell phone is ready for use in almost any place and at any time, which is to say it is always available for its consumption. It is also safe to use, there is not much perceived danger in using it[24] – not like water heaters that used to explode, or automobiles whose use takes hundreds of lives every year in Turkey. Furthermore, it largely conceals its production process, for the labour and the scientific discoveries that enable it to function as a technology are subsumed in its machinery. Also people tend to change or upgrade their cell phones to new and different models, which provide more varieties of use or simply 'reflect' their taste more accurately.

Yet the cell phone is not a totally inaccessible device. One of the ways in which we can consider an object as a device, as Borgmann suggests, is to consider the requirements for repair when the object is broken or becomes obsolete – whether or not the technological object is a throwaway item. Kamil has told me that now that the cell phone has become the 'king' technology of our times, people are much more familiar with its mechanical aspects. He says, 'some even know how to fix a broken cellular better than me', and he adds, 'you don't need to have qualifications for this, you can just check the websites and learn how to fix the minor problems'. He himself became a repairman largely by tampering with the machine. However, the fact that some users can fix and repair minor cell phone problems does not mean that its technology is fully transparent and accessible; if people cannot fix their broken or obsolete cell phone, they go to a shop like Kamil's to have their machines repaired. In Turkey the cell phone is not a throwaway item. If one's cell phone is broken or become obsolete, one does not immediately replace it with another that offers the same functions or similar meanings.[25] Rather, one first tries to have it repaired, and if that is not possible then after 'transferring the personal stuff' into a different device, one purchases another cell phone. One of my interviewees told me:

> My cellular was once broken. It had just started working occasionally. Normally when an object is broken that you use on a daily basis, you just throw it away and get a new one. Because you need it all the time. But with a cellular, it is not that easy to throw it away, because it is not that cheap and also you have lots of things in it, like telephone numbers, nice messages, your photos and so many things like this…but it's also hard to get it repaired, because you need to wait at least couple of days for its repair. So the best thing to do is to have a spare cellular which you can use when your main one is getting repaired. (Kaya)[26]

In addition to facilitating access to its technical functioning, and not being easy to throw away, the cell phone allows and even demands its users' involvement and bodily engagement for its functioning. Thus, the cell phone cannot be straightforwardly considered as a device (as Borgmann defines the term) which prevents human engagement with its machinery and which functions as an easy throwaway item. In order to make the cell phone work, one needs to open and close the machine, charge the battery when it is empty, pay attention to it when it rings or vibrates, and get involved in its machinery to take or refuse a call. User involvement is also necessary for other functions like writing and reading text messages, recording and playing videos, or downloading and listening to music. In this sense the cell phone is not like a central heating system, with which one has to do relatively little to be a consumer of heat. The cell phone does not make human involvement impossible, on the contrary it makes involvement simpler. One does not need to be qualified or be skilled in order to take a picture, write a message, make a call, or play a game.[27] Cell phones are designed for unskilled users who like to be involved in their essential objects' machinery. In other words, it is designed to give the unskilled user a feeling that s/he can control the machine. After all, the machine's function is dependent on the hand's direction.

The bodily habit of using a cell phone refers to the ongoing adjustment of mobility: 'we take the motor space of our interaction to our hands, to the space of our bodies and come to know its model-specific characteristics in the same way that we know the placement of our own limbs and fingers' (Richardson, 2007: 209). One develops a haptic familiarity with the switchboard or touch-tone keys on the screen and gains the ability to use the phone with one hand, whether texting, speaking, recording or playing. The cellular is perfectly incorporated by the hand, entering into an intimate and habitual relationship with the body (see also Oksman and Rautuaninen, 2003). Amparo Lasen argues that it is 'the way mobile phones are held and touched that make this relationship different than other ICT devices. The attachment to mobile phones is revealed by the transformations from being an object always at hand to being almost always in the hand and close to the body' (2004). As Adrian de Souza e Silva (2007) comments, among all mobile technologies, cell phones are the closest to the body. The cell phone offers visual, haptic and acoustic experiences and 'demands variable and oscillating modes of somatic involvement' (Richardson, 2007: 209). Cell phones include a space for users to input and store personal information, thereby customizing their machines. Different types of data – images, sounds, videos and games – can be downloaded into a cell phone; users are encouraged to be involved in their technological object's inner and outer space, to make it their own, to embed it in their everyday lives.

Thus, the cell phone is too personal to be considered as just a device, which, as Borgmann understands the term, functions merely as a commodity. People personalize their cell phones in a number of ways, and the most significant and affective way of making the machine one's own is to put things in it that make self-image or a desired self-image possible. In opposition to Borgmann, Verbeek concludes that 'attachment with the product can arise only when the machinery of the product makes involvement possible' (2005: 227). I suggest that while the cell phone is clearly a commodity, and that its commodity aspect indeed makes it more addictive, attachment and addiction cannot be understood solely from the perspective of consumption. The cell phone's commodity aspect is only one dimension of the addiction. As I mentioned at the beginning of this chapter, I am not talking about an addiction that has no ties to the attachment. Specifically, the cell phone is not merely a drug or intoxicating commodity, only giving the consumptive pleasure of an hallucinated exteriority free of one's preparation, choices and purposes. Rather, one is involved in preparing her/his own drug, one participates in producing one's own commodity.

In an effort to understand the collective attachment to cellular telephony, we need to pay attention to the affective, sensory, imaginative and psychic relations between the artefact and the body. An attached artefact for collectives, cellular telephony has its own weight, as Don Ihde (1990) writes. It opens up a space for human involvement, producing various effects that make itself an attachable artefact. It is because the cell phone plays a mediating role in people's relations both with themselves and with others that it is more than a commodity; it is not only a machine but a technological performance of material and symbolic meanings and bodily engagements (see chapter one for the discussion of performance). In that respect, the attachment I am identifying is not an emotional attachment that 'arises out of the memories with which they are associated or their family references' (Verbeek, 2005: 233), nor is it just an attachment to the commodity. This is an attachment to the cell phone's performance, which incorporates commodification, individual meanings, involvement, and bodily engagement, and thereby binds imaginations, inclinations, responses, and affectivities through its performance.

Cellular telephony: full penetration to life

> I don't see it like a machine, not like a technology, it has become everything to me, like my family, my life...(Meryem)

> [The cell phone] is a gadget without which you feel you have no connection to life. (from the website of ekşisözlük)[28]

> Connect to life with Turkcell, with Turkcell connect to life. (Turkcell 2007 ad slogan)

When searching for cell phone stories in the media, one of the themes that caught my attention is the popularity and news value of incidents that bring the cell phone and death together in the realm of the everyday. As one scans the news, one senses that the dying or dead body's connection with the cell phone is expected to provoke surprise and shock. The journalistic narration of the cell phone tells us that this technology is a 'life' technology, not only in terms of preserving one's life and lived experiences in its virtual space but also as it appears to breathe energy and life into its user. News programs, whose discourses are often dominated by the stories of violence and death, will assign greater news value to an ordinary traffic accident when the victim's cellular is ringing at the crash site, as if this is supposed to make us think that the victim died despite this technology. There is the whisper of a fearful truth, 'technology will still let you die',[29] in the irony and absurdity of a cellular as a synonym of life ringing on the body of the dead.[30]

The representation of the cell phone as a life technology is not just a surreal or hallucinatory articulation; it is based in part on life-saving experiences. The people of Turkey dealt with two destructive and deadly earthquakes in 1999 and 2001, in the country's northeast region. Those that had their cell phones when buildings collapsed, trapping them under piles of rubble, were able to call for help. We have seen, read and heard a number of stories in which the cell phone has literally saved lives by enabling people to connect with each other. This has reinforced the association of cellular telephony with life, especially for people who live in high-risk earthquake zones. In some of these accounts cellular telephony is also praised for taking on the role of state organizations that are normally expected to find living bodies under collapsed buildings and carry out rescue operations (see for instance, Civaoğlu, 2001).

The cell phone becomes associated with life when life is perceived to be at risk due to a lack of social order – the increasing street violence, political terror, the perceived threat of robbery or rape – or the absence of an institution that is supposed to care for its citizens' lives. As I note in the first chapter, there are indeed some researchers who argue that cellular telephony functions as a security blanket, especially for women, children, and urban dwellers (see for instance Katz and Aakhus, 2002; Garcia-Montes, 2006). Yet, in my own observations I have noticed a more ambivalent interpretation about the security the cell phone ostensibly provides. Some of my informants have said that the cell phone does not increase security, rather it attracts attention and makes people who happen to use it in public spaces more vulnerable.[31] A taxi driver told me that he often hides his cell phone, or just puts it in his pocket while driving clients to their destination, because 'there are people who can kill me just to get this phone and sell it in the second-hand market' (Ahmet).[32] A young man who often hangs out at night told me that he tries not to use his cellular in public spaces so as not to invite possible robbery (Emre).[33] A transvestite nightclub dancer, who says she is often harassed by men in the streets of Istanbul, believes that a cell phone cannot protect anybody who is in real, concrete danger (Nilüfer).[34]

Yet others (especially my female interviewees) expressed how safe and secure they feel when they have their cellular with them. Emine told me that when she takes a taxi late at nights, she often prepares her cellular to call a friend in case anything threatening happens. Another said that she sometimes makes fake calls to an imaginary police officer when she is surrounded by some men who appear menacing to her, and that this 'is one of the ways that I have found to protect myself' (Ilknur).[35] A young university student who moved from a village to Ankara for her education expressed the sense of safety she feels with her cellular as: 'At times when I think I am lost in this big city, feeling my cellular in my hand gives me the feeling that I'm still safe. I can call some friends or my aunt to ask where I should go or to pick me up from where I am' (Yeşim).[36] In a somewhat forceful manner, a middle-aged single mother of two grown-up sons told me: 'My cell phone is like a head of the family. When I have my cellular, I feel like I don't need anything else, I can be broke, I can be all alone yet still I am safe and protected' (Kamile).[37] Such statements testify to how the cell phone affects women in Turkey, so that its nearness to the body instigates and facilitates forms of movement for women in urban spaces where previously such mobility had been limited. By being close to the body, it gives a sense of safety to women, for whom public spaces mean more risks, threats and danger.

The experience of feeling safe and secure with a cell phone also points to its affective and sensory dimensions. Emphasizing the cell phone's affective sense of safety and security is not to imply that its actual functions do not offer any measure of safety. Instead it is to argue that affective and sensory experiences work in concert with instrumental capacities to provide this feeling of protection. Although most of the time, one is consciously aware of the fact that the cell phone might not work when one desperately needs its connection, or that using a cell phone might be too late in the face of any possible threat, feeling a cell phone in one's hand or close to the body still gives this sense that one is safe. It is an affective and psychic protector, a talisman with which one feels secure, still connected even when alone and vulnerable to the hazards of the city. As Plant writes: 'I'm physically alone, but not isolated and alone, because I'm still embedded in my social setting' (2000).

The associations between cellular telephony and life similarly refer to this affective, sensory and unconscious experience of cellular telephony, to how its presence and nearness gives the sensation that one is connected to life. Although people who feel and treat this technology as a life-technology know very well that life is possible without a cell phone, they feel that they are missing something crucial when they do not have their cell phone. In my interviews with people of different occupations and ages, residing in different residential districts or in different cities, most told me that they feel a sort of discomfort, anxiety, fear or even guilt when they do not have their cellular close to their bodies. It seems that the cell phone has diffused into life in such a way that operating without it becomes unthinkable, unimaginable in people's life-world.

Hyperbolically, we talk about an attachment when we miss an artefact; we feel depressed and empty, as if our bodies are disconnected from life. More typically, however, we speak of discomfort and anxiety when the cellular is not close to the body. Yet this is not expressed as an emotional attachment, a specific 'love' of the machine/object (although this can be the case for some, particularly if a machine symbolizes and carries certain memories and personal associations).[38] As a general tendency, we simply speak of this attachment as inevitable and necessary, as if cellular telephony attaches us to life itself. There is almost a physical need to be with it, to care for and give time to it. Kaya's answer to my question about the incitement to respond to a ringing cellular is illustrative. For him 'the call' of the cell phone is analogous to bodily excretions whose intensity urges one to move and act according to its orders. Kaya says:

> It's like the feeling [sic] an urge to urinate…like as if your bladder is full, and you have to empty it, right away. In the same way

> I respond to my ringing cellular…or it is even worse than that because, for instance, I can't pee on public transportation, but I answer my cellular right away, regardless of where I am, who I am with. (Kaya)[39]

One woman who sleeps with her cell phone under her pillow tells me:

> I am a slave to this machine, totally attached to it. But, for instance, not in the way that I used to be attached to my walkman, my walkman was something that I loved. Yet my cellular is something that is necessary for my life. Not perhaps like I can't live without it, but in some ways yes, I can't live without it. A number of times I went back home and got a taxi to get my cellular because I just feel so unconnected…Now I don't leave it anywhere anymore. Wherever I go, it is with me… even at nights I am with it…I sleep with my cellular in my bed [she laughs]…can you imagine I am sleeping with a machine instead of a lover or husband? [she laughs again]…as a matter of fact, it is a disease! (Gaye)[40]

The 'disease' manifests as an addiction or compulsive attachment to cellular telephony, without which 'life' becomes unimaginable. So portable that it can be carried almost any place the body goes, accompanying the body in almost all its movements, evading the difference between indoor and outdoor and, most importantly, being a possession that creates a sense of mastery, the cell phone becomes associated with one's life and individuality, including one's secrets.

Handling the life technology: bodily incorporation of the cell phone

> It has become part of our bodies, like our hands, our arms. (Gülden)[41]

The particular and distinctive habitus that has developed around the cell phone is primarily based on the ways in which the body relates to the machine. As I note in the foregoing discussion of bodily involvement, the cell phone is incorporated first by the hand. After suggesting that cellular telephony has become associated with life itself, rather than social status or lifestyles in contemporary Turkey, I would like to re-visit the work of the hand in generating this association between cellular telephony and life.

Many authors have conceptualized communication technologies as prostheses for the body (for discussions about the cell phone as a prosthesis see, Oksman and Rautuaninen, 2003). Understanding the cell phone as atechnological prosthesis means extending its reach to the entire body. The cell phone does not act as a prosthesis for any one body part; it speaks to the ear, it is spoken to through the mouth, it has a screen which becomes larger and larger as the machine becomes smaller and thinner. In this respect, it does not function by mimicking an organ's work or by replacing an organ. Rather, it does the work of most of the senses, and so demands variable and oscillating modes of somatic involvement (Richardson, 2007: 209). Yet the hand is always needed for it to function. For Richardson, cellular telephony refers to 'the finely coupling hand and mobile device' (2007: 209). Cellular telephony needs a hand to operate, grasping it for the body. The body comes to know the cell phone just as it knows its limbs or fingers through the tactile clutch of hands. As the grasping organ of mankind the hand has been questioned, philosophized in various ways. So has our tactile relation to technology. We were told that hands are for mastery or that hands are lost, replaced,[42] amputated by new technologies (Kittler, 1999; Seltzer, 1992).

In the case of the hands incorporation of cellular telephony, I suggest that the hand does not simply vanish to be replaced by the machine, rather it moves back and forth between feelings of mastery and alienation. The work of the hand, Mark Seltzer's theoretical work suggests, necessarily intimates the translation from inward to outward, from the psychic and spiritual to physical and cultural. Translation is one of the core concepts of my next chapter, framing my discussion of the inner logic of cellular telephony. When we speak of cellular telephony's incorporation by the hand, and of its work as translation, we can consider the hand not as lost or replaced, but as a means of transformation, possibilities, interruptions and articulations. The hand opens the body not only to itself, but also to the world.[43] Therefore the hand is a site of potential, of transformation through contact.

Handling this technology that has come to be associated with life might provide the sense that life can be controlled, ordered and organized; noticing the hand's addiction to the machine might generate the uneasy feeling that what is presumed to be controlled is actually controlling the body. As a site of possibilities, the cell phone is an object of the desire for control and regularity; it also functions like a living body to which the hand yields control over the body's actions and movements (see Sofia, 1999).

Bodies at times feel controlled or mastered by the cell phone. Gaye was not the only one to express discomfort at being addicted to this machine; many others told me that they often feel like slaves to their machines, that they automatically answer their cellular's call. Although the machine positions itself as user of the body, bodies are still attached to and control cellular telephony through their hands. I suggest it is because cellular telephony promises multiple possibilities of lives that the hands of melancholic bodies are always already in search of 'objects'[44] that can provide this sense of contact, transformation, and articulation. In the next chapter, I explain ways of looking at cell phone performance as a pocket technoscape, one that can be handled and is incorporated by hand. This approach emphasizes potential reasons why this technology becomes an addictive performance in Turkey, and opens up ways of exploring how cellular telephony's modes of containment are linked to the social practice of cellular telephony in Turkey.

5

INDIVIDUAL ARTICULATION WITH CELLULAR TELEPHONY: CONTAINMENT, TRANSFERENCE AND TRANSLATION

> It contains all of my memories, my photos, messages, telephone numbers of my friends…it is like me...one can tell what kind of person I am by looking inside my cell…but I don't like being dependent on it, checking it all the time, organizing all the photograph files, SMS files…spending all my spare time in naming these files and listing them…yet I cannot renounce it…I detest always expecting someone to call me…sometimes when my phone is ringing I try to resist answering the call or checking who is at the other end… I just wait for a couple of seconds to see whether I can stand it ringing and not responding, but finally I give up and respond. (Eda)

> Yes, because you wonder who is calling, if s/he has something important to say…as if you wait for something very important to happen…I even get excited when my phone rings…but for example not in the same way that I am excited when I see my boyfriend or someone that I am attracted to...a weird excitement. (Merve)

As discussed in the previous chapter, notions of the cell phone's meaning or its instrumental value do not provide a comprehensive account of why Turkey's attachment to cellular telephony is so intensely affective. The collective attachment to cellular telephony does not come to the fore as 'love', but rather as an addiction to its affective and imaginational experiences. Sensations, inclinations, wishes, desires, and conscious and unconscious purposes need to be taken into consideration in order to come closer to understanding how bodies that hold different demographic characteristics become attached to this technology. In this chapter, I investigate psychoanalytic[1] interpretations of such attachment in my attempt to 'contemplate' the invisible work of cellular telephony.

In this chapter I focus on two interrelated issues – containment and translation – which come to the fore in making and articulating individuality. These two core analytical domains offer ways of approaching how the cell phone moves people, what it promises them, and how it gathers people together. In the first quotation above, Eda speaks of her telephone as something that contains her life and her personality, by carrying pieces of her memories, her social network and the outcomes of her interaction with others. It functions as a container while Eda sees her presence as contained within this attachment. Yet she is not entirely comfortable with this, like many others who are aware of the controlling force of this technology. When Eda was speaking about the impossibility of ignoring the cell phone and its call, Merve interrupted and gave a reason (an excuse) for why she is addicted, like Merve herself, to this artefact. The sense of possibilities, of an encounter with something different, the potentiality of an event that could be important connects Merve to the cellular telephony. She gets excited when her cellular rings, yet not in the same way as when she sees her boyfriend or someone else to whom she is emotionally attached. In my analysis, I consider this bodily sensation of possibilities in relation to the cell phone's work as translation (and transference). In what follows, I clarify the internal relation between containment and translation or transference within cell phone performance, particularly from the standpoint of a projective identification and an articulation of individuality.

Understanding technologies' containment dimensions

Technologies as containers, and the possibility of projective identification with the technological artefact, have been discussed by the renowned feminist philosopher of technology, Zoe Sofia. She (2000) showed us that technologies are not just machines or tools, that their containing dimensions enable the machine or tool to function or exist as what it is. Beyond machines and tools, she writes, there are technologies of containment and supply, which she categorizes as utensils (like baskets or pots), apparatus (the specialized container in which something may be created or transformed), utilities (such as reservoirs, roads, or buildings) and modern power (e.g. electricity). By emphasizing the aspects of containment and supply, Sofia reveals the internal connections between machines/tools, 'containers', and socio-technical as well as physical environments. Machines or tools come to be seen as contained objects, and so need to be examined within the macro context of containment. This context includes socio-technical realities, and the conscious and unconscious relations of human beings as well as non-human entities. Along with stressing the containment dimension of all machines and tools, Sofia provides multiple examples of container technologies which are also hybrid machine-containers. She writes:

> [T]his category includes a huge range of technologies and relations, from intimately wearable containers like clothes, shoes or condoms to walk-in partly-automated spaces like houses, cinemas, shopping malls, or cities, floating and submersible containers like boats or submarines, nuclear containment vessels, as well as the virtual worlds of computer and video games. (2000: 187)

Sofia (2000) suggests that container technologies offer more than just spatial containment, for they can also house and keep things temporally; the memory of some container technologies provides a space where one keeps things for a certain period of time. Containing in this sense does not only refer to spatial housing but also temporal preservation. This should not be interpreted as a passive function of spaces;[2] on the contrary, Sofia argues, 'containing is to be thought of as a form of action in itself' (190). From this perspective, container functions are integral to the performance of technologies as well. While her work richly discusses the containment dimension of technologies in a number of ways,[3] I will highlight her most relevant ideas about container technologies in order to map the contours of cellular telephony performance for the individual. In what follows, I incorporate her typology of three dimensions to designate the analytical domains through which we can situate cellular telephony as a hybrid machine/container technology. The first dimension addresses how containing technologies might be understood as 'gathering' various elements together to create a certain closeness in performance. The second explains how containing technologies facilitate projective identifications through storage, by housing personal material. Finally the third explores the ways in which containing technologies provide the potential space for a safe means of discovery and invention in the project of self-making.

Drawing on arguments from Heidegger, Sofia argues that containing technologies are 'a gathering together of many elements, forces, purposes and dimensions, both human and extra-human' (2000: 193). A containing technology takes and keeps what it receives, it gathers its supply and 'outpours'[4] what it contains – Heidegger's definition of gathering. For Sofia, containing technologies can be interpreted as technologies of re-sourcing: they can be filled from a source, and they then become the source of what they have kept and preserved (ibid. 192). However, containing technologies do not simply represent such a gathering (or a source); they 'only exist as that gathering of materials, that particular location and shaping and conjunction of space(s), that historical and cultural set of projects and purposes [which they] serve and of which they are an outcome' (ibid. 193). Thus, containing technologies do not function as representational signs or tokens but exist as 'gathering', as an articulation[5] of different elements, whose connection to and roots in historical, cultural and social projects manifest themselves in performance.

Sofia writes that in 'gathering' the key motif is 'emergence': the containing technology 'emerges in a "nearness" or rather a process of "nearing" that gathers remote elements into itself; thus a local and specific object is also a manifestation of its macro-context' (ibid. 194). Containing technologies are themselves spaces which house and store various elements. Within this space, as we will see below, facets of the individual body stand alongside facets of the social and historical; the work of gathering connects one to the others. Containing in this regard can easily be understood in relation to social, historical, psychic, physical and affective performance. This brings us to the second dimension of Sofia's discussion of containment.

Following the premises of intersubjectivist psychoanalysis, Sofia's own insights concern the dialectical relation between containing technologies and their resources, as well as that of mother and infant. Within this discussion the ways in which containing technologies function for the projective identification of the body become clearer. Psychoanalytic accounts of individual subjectivity take the primary relation between infant and mother as their core analytical domain for understanding how an infant becomes a subject. Intersubjectivist psychoanalysis (specifically Winnicottian theory),[6] as Sofia reminds us, suggests that there is no infant apart from its maternal provision. Using the clinical psychoanalyst and theorist Thomas Ogden's review of Winnicottian subject/object relationships, Sofia shows that an infant needs to have a dynamic relationship of containment with its mother in order to become a subject. Considering the womb as the primal container, she argues that an infant's emergence rises out of its containment in the womb, and in return the mother becomes contained in the infant's personality (2000: 183-4). There is thus a dialectical containing relationship between mother and infant, as the mother who had previously contained the infant is necessarily contained in the infant's personality in order for her/him to feel her/his personality (184).[7] As Ogden describes it:

> Paradoxically, the subjectivity of the individual presupposes the existence of two subjects who together create an intersubjectivity

> through which the infant is created as an individual subject. The infant as subject is present from the beginning although the subjectivity exists largely within the context of the psychological-interpersonal (containing/contained) dimension of the relationship of the infant and mother.' (Ogden, 1992: 619 cited in Sofia, 2000: 184)

Ogden defines this relationship as one of 'projective identification', in which 'aspects of the self are not simply projected onto the psychic representation of the object (as in projection), but "into" the object' (Ogden, 1992: 617 cited in Sofia, 2000: 184). In intersubjectivist psychoanalysis the relation between subject and object is the key motif. Although Sofia does not undertake a detailed discussion of how this dynamic relates to that between containing technologies and individuals, she explores one of the core concepts of Winnicottian theory of individuation of the subject: transitional phenomena and the transitional object's possible connection to containing technologies. She writes:

> Other spaces experienced by the infant are the inner world of fantasy and outer world of socio-technical reality, bits of which become caught up as 'transitional objects' in a third space, called 'potential space' by Winnicott (1971). Being able to play safely in this potential space, negotiating between inner and outer worlds and self and (m)other, is an essential part of infantile development, and moreover, Winnicott argues, it is the foundation for later creative experiencing and cultural production which plays (or works) on the borders of fantasy and reality. (184)

Accordingly, the self 'is understood as an entity given shape through various dynamic relationships of containment that both construct and occur in spaces that are interpersonal, imaginative, real, active, the products of conscious efforts as well as unconscious or automatic labours' (Sofia 184-5). By proposing connections between 'transitional object' and 'tool', between potential space and studio or workshop where discovery and invention take place, and between maternal provision and the containing dimension of technologies, which allow the infant and technology (thing) to emerge/be, Sofia constructs an analysis of containing technologies through an analogy to maternal containment. In other words, the containing dimensions of technologies are essentially maternal dimensions, which are generally neglected in the philosophy of technology and become unacknowledgable in the everyday use of technologies. Yet this containing dimension of technologies is a necessary precondition for the becoming of a 'thing', just as maternal containing is vital for an infant's becoming. She suggests that the container corresponds to the mother as 'our first sheltering container and source of supply', into which 'we dump our excess stuff and [that] we come to consider as an extension of ourselves' (187). By likening technologies to maternal containment, as both gather various elements together and 'outpour' what is contained, containing technologies also offer a sense of 'life giving' activities to their users (192). Containing technologies' maternal provisions generate a sort of containment which does not only house what we put into them but also contain what we cannot – the excess.

Similarly, containing technologies constitute (or re-constitute) potential spaces in which various discoveries take place. For instance, a containing technology like an oven, which Sofia defines as an apparatus or specialized container, provides a place where something may be created or transformed. In that sense containing technologies generate an imaginative opening – play space for the infant – where inner and outer worlds are negotiated (189).[8] Importantly, if the relation between body and containing technologies is considered as analogous to the relation between mother and infant, then there needs to be a bond between technology and human body just as there is between infant and mother. The mother ideally identifies with the infant (e.g. anticipates her/his needs) and yet remains 'separate to serve as the container and interpreter of its experiences' (Sofia 184). Winnicott describes this synchronized response as a primary maternal preoccupation (1971). The technological equivalence to this primary maternal preoccupation concerns the degree to which the container adapts to us; the more a technological object responds to our wishes and calls, the more 'user-friendly' it seems (Sofia, 2000: 187).

From the standpoint of intersubjectivist psychoanalysis, containing technologies are also transitional objects, which might explicate why technologies partially originate in the inner world or are always meaningful insofar as they become part of a narrative of human purpose that is a necessary aspect of one's becoming (Sofia, 2000: 189). When all these issues are thought together, they enable us to consider containing technologies not only as objects with which we identify but also as spaces where aspects of our selves and our self-narrative are contained – spaces in which we 'dwell'. Sofia argues that containing technologies provide potential spaces for bodies; bodies discover things and investigate their own feelings in this containment, and thereby identify with them (containing technologies). Housing the human and non-human entities, containing technologies also become laboratory spaces where one experiments with one's own life. Containing technologies allow for and encourage the storage of one's own discovery of self, including failed attempts at imagining proper self-inventions, and thereby become objects of projective identification. This is significant for my contention about cellular telephony being 'instrumental' in articulating individuality and individual spaces/worlds. However Sofia's analysis does not delve into the ways in which containing technologies open up a space for the human body to find the means of articulating a self through play in this space. In the next section, I will attempt to work through these issues by invoking, among others, Avital Ronell's theory of telephony/technology.

The third issue that I would like to highlight here is the safety of the potential space where discoveries are imagined or actually take place. As mentioned above, Sofia tells us that containing technologies 'constitute (or co-constitute) environments and locations in themselves' (192). People dwell in these spaces, much as they reside and remain in buildings. Such spaces are a 'safe preserve' for bodies, as the sense of dwelling is a 'basic, habitual and inhabited condition of human life' (Heidegger, 1971b: 146 cited in Sofia, 2000: 192). Sofia writes that this 'safe preserve' of containing technologies that exists between fantasy and reality corresponds to 'maternal functions as one of actively containing an emergent subject and letting it play in potential space, so it can become who it is' (193). Transitional objects encourage safe play in the potential space, and in negotiation between self and (m)other. As potential space, a containing technology becomes 'safe' because, like a mother, it 'works unobtrusively to ensure "smooth functioning" and continued supply to the infant whose bodily states and feelings, she regulates' (189). The notion of 'safe play' suggests yielding control to containing technologies, allowing them to regulate actions, bodily states and feelings in ways that correspond to the maternal containing function.

To understand how the cell phone works to articulate individuality, we need to extend Sofia's arguments about the possible connection between containing technologies and maternal containing functions, and elaborate on how an individual 'becomes' through relations with the transitional object and/or (m)other. Specifically, we need to work more on the imagined 'possibilities' enabled and encouraged by the technological-transitional object. Although Sofia's discussion of the connection between containing technologies and projective identification does not invoke the concepts of translation and transference, they are intrinsic to intersubjectivist psychoanalysis, which has inspired her analysis of containing technologies. In what follows, I will expand upon Sofia's approach in order to address how transference is intrinsic to the psychic process of loading a tool with the body's inner worlds through the body's use of or play with the tool. Avital Ronell's approach to how these processes work through the telephone will shed light on our examination of transference/translation and technology.

Psychic individuation, transitional objects and transference

In *The Telephone Book: Technology, Schizophrenia and Electric Speech* (1989), Avital Ronell writes that the telephone is associated with a maternalized force (20-22). Like Sofia, Ronell suggests that just as the maternal figure is withdrawn as a child's individuality broaches the psychic domain, in its use as a sensory object the telephone's maternal connotation sinks away and becomes unrecognizable, yet its presence remains felt. Paralleling Sofia's argument that a tool's containing dimensions echo maternal containment, Ronell notes that the telephone is characterized by an intrinsic maternalizing call. The telephone is both lesser and greater than itself – lesser when it is considered solely in terms of its material functioning, and yet greater in terms of a maternalizing call which beckons as a home (the womb) that is lost or absent (ibid. 20-21). In this respect, the telephone conveys a sense of not-being-at-home. The child is departed and yet desires to return; this is why s/he takes the telephone's maternalizing call.

This maternalizing call can be interpreted as corresponding to that maternal provision apart from which the infant cannot be who/what it is, as we have discussed above. Ronell writes, 'the bildopedic culture has produced itself out of a combinatory lack and excess; it constellates the human subject telephonically' (1989: 20). Just as transitional objects (tools) facilitate the projective identification needed for an infant's individuation, the telephone functions for child's subjectifying relation with his (m)other and/or Other.[9] Situating the telephone as a tool that contains – or, more precisely, preserves – the absent maternal figure also addresses the psychic work of transference, which transforms a material object into an object that contains, among other things, the maternal image. Working simultaneously with Sofia and Ronell, we shall dissect the concept of transference through referring to the works of Ogden and Freud, which have been the inspiration for Sofia and Ronell respectively.

Thomas Ogden (1992) describes transference as the externalization of an internal object of the self into another object, which then comes to the fore as an external transitional object with which the infant identifies. 'The outcome of this 'internalization' of an object relationship 'necessarily involves a splitting of the ego into parts that, when repressed, constitute internal objects which stand in a particular unconscious relationship to one and another' (Ogden 1986: 131-132). Through transferring and externalizing the internal object's unconscious thoughts and feelings for the (m) other to the transitional object with which the infant has projective identification, the infant becomes what it is. As mentioned above, Winnicott (1971) coined the term transitional object or transitional phenomena to explicate the process of the infant becoming an individual or the infant's growth through her/his relationship with objects. Winnicott refers specifically to objects such as a blanket or a teddy bear, which become psychic representations of the mother holding the infant (simple projection) in the absence of the mother. This serves as an 'illusionary experience' for the infant who discovers and creates objects to replace maternal care. In the absence of her/his mother, the infant becomes capable of being alone by internalizing the maternal environment, to freeze an emotional situation in time (Ogden, 2004: 1353). Ogden writes: 'It is the function of the mother as holding environment (as opposed to the mother as holding object) that is in the process of being taken over by the infant or child' (ibid.). In a depressive position, i.e. the child cannot tolerate the fear evoked by the absence of his mother, 'holding sustains the individual's experience of a form of being that is continuously transforming itself – an experience of remaining oneself over time and emotional flux in the act of becoming oneself in a form previously unknown, but somehow vaguely sensed' (ibid. 1354).

This unconscious transference of the internal object into the transitional object is one dimension of psychological growth, while the dynamic relationship of container and contained is another dimension (Ogden, 2004). These two dimensions are not the same, rather they designate different vantage points of the same process, that of projective identification (ibid. 1349). For Ogden, 'the idea of the contained and container addresses not what we think, but the way we think, that is, how we process lived experience and what occurs psychically when we are unable to do psychological work with that experience' (ibid. 1355). While 'container' refers to the capacity for dreaming and thinking, 'contained' refers to the unconscious thoughts and feelings that derive from lived emotional experiences.[10] Through recounting his clinical experiences, Ogden shows us that for the dynamic relationship between container and contained to take place, intersubjective relations (connections between internal objects – i.e. the swallowed (m)other – that are contained in each subject) and interpersonal relations (pressure on the other (person) to act in accordance with the fantasy – i.e. no absence of mother, no loss of self, no pain) are ultimately conditional (2004).

Projective identification, then, is a nodal point between intrapsychic (transference), interpersonal and intersubjective relations, and thereby becomes the condition for the infant to become an individual. Without projective identification, the individual would remain locked in a solipsistic world and would fail to develop as a human being (Ogden, 1989).

Ogden thinks of projective identification as an unconscious process which involves a fantasy of placing one's own aspects (like internal objects, along with feelings/thoughts) in another, who is thereby controlled from within (i.e. transference, through which the mother becomes the other, or an object). Interpersonal pressure on the other to think, feel and behave in accordance with the fantasy (the mother will respond, or the object will respond accordingly) ensures the return of the sent aspects (like internal objects along feelings/thoughts) in an altered form. Ogden maintains that the process of a projective identification by the other (recipient) 'is not simply a matter of returning modified psychological contents to the projector', instead he says, 'it is a matter of altering the intersubjective mode of containment generating a new way of experiencing the old psychological contents' (1989, 130, my emphasis). Thus the relation with the other is conditioned by a desire for changing what is at hand, transforming what is already being experienced. This is a significant point – which will become clearer in the forthcoming sections – for our discussion of cellular telephony as a containing technology which contains aspects of the individual, including her/his desires, wishes, conscious and unconscious purposes, inclinations and imaginations.

Having discussed the ways in which transference alongside contained/container dimensions is at work within the process of projective identification that entails intrapsychic, interpersonal, intersubjective relations, we shall now return to the discussion of the transitional object that contains bits of inner and outer worlds in the third space – that is, potential space which unfolds as a discovery space for transforming the old psychological – emotional – contents. In the case of technological tools, the play with the transitional object is necessarily continuous and repetitive. From a Winnicottian understanding of individual articulation, the transitional phenomenon is always already crucial, a condition for negotiating and stabilizing individual identity. The transitional object's instrumentality offers an action and a movement, opening a site or potential space where the infant/child relentlessly tries to discover, to invent, to arrive at the imagined proper individual identity. Freud's (1990) speculation about his grandson's play with a woollen reel, the movement of sending the reel out of sight and pulling it back by its string, illuminates the kinds of movement the transitional object provides for the project of inventing some other ways of altering the already existing psychic contents.

In *Beyond the Pleasure Principle*, Freud (1920, 1990) talks about his grandson's game of *fort/da*[11] as an exemplar of transference.[12] In recounting the game, Freud emphasizes that the child's mother is not there, and Freud assumes the child suffers the absence of his mother as a loss. Freud's interpretation of this game is that in his mother's absence the child has transferred his feelings onto the reel; by sending the reel – or substitute mother – out of sight and bringing it back the child develops a ritual of mastery. In Ronell's (1989) reading, the reel is not a psychic representation of the mother, instead it is simultaneously a thing, a subject, and an object which preserves bits of the mother and of the boy, who can 'be' through the movement of disconnecting and then reconnecting with the (m)other in new ways that would ideally take him to his self-proper. The boy sends the reel away (his mother and himself), thus risking his presence along with his mother's. He masters himself/her by holding the string, and makes himself reappear by pulling the reel back. Ronell writes:

> Disappearing like her and making her return along with himself, the child identifies with the long-distance mother. Effecting his own disappearance, he masters himself symbolically and he makes himself reappear in his very disappearance…keeping himself (like his mother) on a string, on the wire. (85)

Despite the threat to self-presence, it is a safe play.[13] The movement of *fort/da* is an imagined arrival at self-presence, which cuts off all losses and absences. The reel is not only transference of the absent mother but also of the loss of self-proper, a translation of absence into presence. As both the condition of possibility for movement and impossibility of arrival, the reel is a catalyst for self-articulation. Insofar as the reel substitutes for the lost womb, the lost perfected shelter in which the child feels at home (mastering himself, mastering life against death (Bronfen, 1989)), Ronell argues s/he will never come home; the reel, the telephone, the courier (see chapter three and Derrida's discussion of the courier) is itself what precludes the arrival. Yet the child will continue to desire to be called by the absent mother, whose image sinks away and is preserved in the string (transference and translation) that makes the movement (between appearance and disappearance) possible. Put in another way, the game is played repetitively by the child, because s/he is transferred into the reel and into the string, and his imagined articulation is transferred into the movement and it is always incomplete.

Just as the play of *fort/da* becomes appealing and addictive insofar as the losses are transferred into the movement, the telephone – a synecdoche of technology, a technological reel – also opens up a space for melancholic bodies to engage in compulsive behaviour, a space where losses are transferred and translated into the movement of calling and receiving the call. Importantly, Ronell notes, the telephone is entrusted with transference and translation (85). It actively takes up the role of the boy who both desires to connect to his (m)other and disconnect from her. Like transference, the telephone instigates a relation with what is absent and like translation it urges a relation with the other through improvisation and endless play. As she writes, the telephone is 'a live' at times, 'or at least "life" punctually gathers in it and takes part in it' (ibid.).

Thus there is an element of the fairy-tale wish in telephonic movement: finding the essential connection that will transport oneself to proper self-presence. The telephone is a response for the 'boy' who is in need of a sort of play to soothe him, to pacify his pain in the absence of a connection to the (m)other. For Ronell, 'to be what it is, it [the telephone] has to be pluralized, multiplied, engaged by another line and heading for you' (1995: 5). In other words, for the telephone to work as maternalizing call, to instigate independence, it needs to connect to the other so that the mother image dissolves into indeterminate They. The telephone, says Ronell, destabilizes the identity of self and other, subject and thing; it is 'unsure of its identity as object, thing, piece of equipment, perlocutionary intensity or artwork; it offers itself as instrument of the destinal alarm, and the disconnecting force of the telephone enables us to establish something like the maternal superego' (9).

As a line between self and other, a force of connection and disconnection, the telephone is characterized by the impossible wish to take the self into self-presence by allowing the body to relate itself to other(s). A proper distance between self and other would ideally bring a proper self-presence. The telephone-as-transitional object is not only illusory in transforming its thingness into a meaningful practice, but also hallucinatory in its unconscious work of cutting all losses and bringing them into presences. Ogden, Sofia, Freud and Ronell's discussions all address crucial aspects of the technological artefact as a transitional object for psychic individuation and for intersubjective and interpersonal relations. Ogden's projective identification along with Sofia's correspondence between transitional object and containing technology provide the means to grasp the work of containing technology in the intrapsychic, interpersonal and intersubjective relations through which the body searches for finding ways to erase the one at hand, alter it, transform it to arrive in the imagined impossible self-presence, the proper, the home. Ronell's critical review of Freud's *fort/da* theory lets us consider how the repetitive and continuous play with the transitional object becomes a sort of 'contract'[14] for the searching of the loss of self-proper. In other words, the loss is integral to the transitional object's performance.

Playing a central role in these other ways of experiencing the existing emotional contents, the transitional object is part of an impossible project, promising a laboratory space where the body can improvise, can try to find a self-proper amongst others. In the case of containing technologies like cellular telephony, if the transitional object enables projective identification by opening a site where relating to others is crucial and necessary, then we can surmise that the transitional object (the cell phone) works for both intrapsychic and interpersonal translation. After all, projective identification or individuation processes cannot occur outside of a relation with others. Relating to others, 'communicating' with others, instigates translation in every moment of one's attempt to understand the other as well as to understand oneself in order to express that to another.

Against the familiar contention that communication is the mutual understanding of different parts, Ted Striphas (2006) suggests that communication is essentially translation. Within the process of communication, one is always confronted with foreignness and otherness that require translation and improvisation. He writes:

> As understood from the standpoint of translation, communication refers to the process by which we first interpolate another's address into, and then interpret it using, our own unique sign systems. The process by no means assures understanding, much less a neatly ordered and organized social world, but provides instead for our everyday ability to improvise imperfectly and haphazardly with others. (2006: 239)

Not only does one's sign system require translation for and by others, but one's own language is also an other for oneself, and therefore needs translation. Translation implies potentialities for articulation, since through improvising one looks for meaning, properness, self-presence or home. In the laboratory of social life one experiments with various sign systems in order to relate to others and to find oneself. Translation and improvisation with others make it possible to search for ways of altering and departing from the self at hand or the time-space where the body dwells in. This is how the technological transitional object can become an attachable object, as it creates a site where translation brings forth new possibilities and a flexibility for improvisation. Interpersonal, intrapsychic and intersubjective translations are inter-relational, they come to be bound together within the process of projective identification. However, a technological artefact's contribution to articulating individuality through projective identification cannot be understood solely as the strict formulation of object relationships within the psychoanalytic model of maternal relations. As Sofia writes, through playing with the transitional object – the technological tool – one negotiates the inner and outer worlds which are embedded in social, historical, and political realities and structures. In previous chapters I discussed the ways in which the social and the historical enter the body, and I designated how the social is also the base of emotions, or more precisely, where melancholia is invested. Thus searching for 'new ways of experiencing the old psychological contents', which Ogden tells us are emotional in essence, does inevitably speak to the possibility of altering melancholia. If the transitional object becomes what it is through the transference of feelings and thoughts about an internal object, and if the psychic domain of an individual body is not outside of the terrain of the social and political power (as I claim in chapter two), then transference clearly has social and political dimensions; just as contained/container are descriptions of how we think, feel and dream, transference is about much more than familial relations.

Following from his contention that the psyche's becoming cannot be reduced to identifications with mommy-daddy figures, Guattari (1995) suggests that the psyche 'engages, assuredly the register of language, but also non-verbal means of communication, relations of architectural space, ethological behaviors, economic status, social relations at all levels, and still more fundamentally ethical and aesthetic aspirations'[15] (Guattari, 1995: 204). Situating subjects of the capitalist world as schizophrenic machines, Guattari suggests that the transference at work in this schizophrenia is 'literally a cry…which interprets alienation, not of [the] schizophrenic himself, but of the others' (91). Schizophrenia is the hysterical symptom of the capitalist body, whose transference betrays not just a culture's political, social and historical desires, but its diseases as well. For Guattari, the schizophrenic transfer is the language of the repressed political.

Jung's recommendations also explicate the social aspects of transference. By divorcing transference from a fixation on familial relations (as in Freud), Jung extends the notion of transference from an individual's recurring and repressed feelings for the maternal figure to the repetitive organization of replacing what is missing in the social. He writes:

> The transference phenomenon is without doubt one of the most important syndromes in the process of individuation; its wealth of meanings goes far beyond mere personal likes and dislikes. By virtue of its collective contents and symbols it transcends the individual personality and extends into the social sphere, reminding us of those higher human relationships which are so painfully absent in our present social order, or rather disorder... What our world lacks is the psychic connection; and no clique, no community of interests, no political party and no State will ever be able to replace it'. (1954: 161, original emphasis)

The umbilical link between the loss of the social and the presence of technologies that are meant to function for connection between different agents becomes clearer in Jung's work. Similarly, Guattari's comments on the political character of transference help us grasp the politics of technologies that are entrusted with the work of translation and transference. Technologies today can be interpreted as powerful performances of sickened mourning for the loss of psychic connection with others (see Ronell, 1989). If, as Ronell maintains, the telephone preserves and swallows what is no longer there, then it also swallows the loss of psychic connection with others. The telephone's translating work thus corresponds to searches for new ways of experiencing sociality and the other, by prompting encounters with an otherness that can bring new possibilities to the articulation of self.

Cellular telephony and the third world 'children' of Turkey

> Wish you had a new cellular in the new year?
> It rings, there is always somebody calling you
> your new messages are good
> in which there are smiling faces.
> I hope you will meet another you in this New Year
> I hope your year will begin in this January
> Connect to life with Turkcell
> With Turkcell connect to life
>
> (jingle from ads for Turkcell, 2007)

Turkcell is currently the market leader GSM provider in Turkey. This campaign, like other Turkcell ads, has been widely seen, broadcasted regularly on all national channels. In its television ads, a blonde-haired and blue-eyed sweet little girl sings this wistful song as a small blonde boy accompanies her on the piano. These two children are avatars of Turkcell technology whose cellular system covers almost the entire country. Turkcell's advertising campaigns humanize technology by representing parts of its technological system as people (who often appear 'extra-human'); this strategy is the company's dominant marketing trope.

Cellular telephony works on radio frequencies, distributed through base stations that cover areas which are divided into cells. In order for a call to be made between two distant agents, they need to be located in range of the base stations' distribution system of radio signals. Turkcell's campaigns represent these cells as 'beautiful' – and mostly blonde – children called *cello-cans*,[16] who appear to do the work in order for subscribers to connect to each other. They are fed milk like many children, yet they are clearly extra-human – they have antennas attached to their costumes. They speak Turkish but do not resemble Turks. Most are blonde, and very western in appearance. They are emotional; not only do they understand us, but they are fully attuned to us: they know what we need, what we feel.[17] They promise a constant connection to our loved ones, they wish new lives for us and they encourage us to meet our new 'selves' (a new 'me') whose possibilities are contained in cellular telephony. They tell us to connect to life, with Turkcell.

When I met with the senior copywriter (Kerem Balcı) of Turkcell ads in Istanbul, we talked at length about their ads, and about cellular telephony in Turkey. He told me that the secret to a successful advertising campaign in Turkey lies in its capacity to trigger melancholy (*hüzün*) and mix it with comedy:

> There are 35 million people using cell phones in Turkey. These are actual consumers, and let's say there are 10-15 million potential costumers. The target audience comes to 40-45 million people. The only thing that would catch all these different people simultaneously is melancholy. But only showing melancholy is not good, it needs to be blended with comedy. People need to smile, if not laugh.

In the recent ad, from which I have translated the above quotation, a *cellocan* tells her wishes for us, that we lose our melancholia by finding a new 'me'. This promise of change and transformation echoes the desire for finding new means of living through the existing psychological contents. The *cellocan* wishes for our cellulars to contain new messages, ones in which there are 'smiling faces' to make us feel cared for and loved by others. The *cellocan* wishes us a new year which would finally be 'our' year with a 'new' cellular. As we have learned to match our melancholic relentlessness with consumer markets, we would of course search for novelty in commodities. This habitus has developed to soothe the discomfort derived from our everyday melancholy (see chapter two). The *cellocan* in this ad – whose appearance contains the imagined western gaze – knows this, and understands us that we need connection: not only to others, but through the detour of others to new 'me's.

The cell phone is a containing technology, gathering together multiple functions of the contemporary technoscape. The simplest cell phone available on the Turkish market contains a telephone, SMS, telephone book, digital calendar and reminders, clock and alarm, call register system, game console, calculator, converter, multiple ring tones and a screen with a little keyboard. The most complex has even more functions, such as an Internet connection, mp3 player, video recorder-player, still camera, location provider, Bluetooth, audio-recorder, radio, and a big screen with a digital keyboard. The cell phone is a containing technology which takes in various forms of 'stuffs' (video, pictures, telephone numbers, et cetera) to gather and keep; its containment makes it a resource for the individual body beyond daily use of the contained stuffs, to work as a site of self-contemplation.

This recalls what Eda told me, that 'one can tell what kind of person I am by looking at inside of my cell…'. Another interviewee, a high school student in Ankara, said 'I often take pictures of my self to see after how I look' (Savaş).[18] Nilüfer, who is part of a transvestite dance show in a night club in Istanbul, mentioned that her cellular is full of pictures of her performances – she looks at them to see how well she danced and how she looked. Other people have told me that they keep messages in their cellular to read over again when they feel lonely, disconnected, 'melancholic' and need to see nice things about themselves. Some said they prefer texting rather than telephonic interaction when they want to convey deep emotions; of love or distaste. However, they keep only the pleasant messages in which there are 'smiling faces'. Others noted that they record anything that seems interesting in their daily lives, only keeping what still makes sense after a while. To my surprise Nilüfer's roommate Banu,[19] who is a transvestite sex worker, said that she never records anything with her cellular; even her address book is empty, she deletes all SMS messages she receives and never keeps track of what she sends to others. When I asked her why, she replied, 'I don't know, I just don't want to have them recorded. I have lots of obsessions, this is one of them I guess'.

The cell phone today works as a 'life recorder',[20] a place where one's autobiographies can be kept in order to generate narcissistic pleasure. The cell phone provides a sense of control by offering options like 'save', 'delete' or 'edit' as one searches for the proper autobiography, for the proper self-narrative. When putting stuff in the cell phone the body holds the string, symbolically mastering her/his own self-narrative. The cell phone thus contains and gathers selective memories and 'outpours' the resource – the assembly of these memories – to the body as a 'gift' that promises 'progress' on the path to proper shelter. In this sense, the cell phone's containment capacity far exceeds those of wallets or handbags (Richardson, 2007). If what could be 'outpoured' did not create narcissistic pleasure, nothing would be put or kept inside.

Even Banu, who does not leave any visible trace in her cellular, tells me that while her cellular does not contain anything about her – there is nothing to reveal her inner world/social world/her – she still feels attached. She does not go anywhere without taking it with her, she feels uncomfortable when she misses it. She tells me that she often calls her mother, her friends and uses it for work.[21] She prefers texting when she has something personal or emotional to say. Yet she never keeps her messages. Cellular telephony promises a containing space for storage, and also a containing performance wherein an endless play of translation, improvisation, and communication with others is possible. As Banu's habits demonstrate, the cell phone's material containing capacity is not the only trait that makes it an attachable technology. Even when it does not literally contain pieces of self, it continues to contain the possibilities for new experiences that can alter the existing psychic, affective, social conditions.

Although technically akin to the wireless radio and functionally resembling a mobile computer, the cell phone has become institutionalized as a telephone[22] that is 'personal', 'portable' and 'pedestrian' (Okabe et al. 2005). It functions as a transitional object for the moving body, as movement here describes travelling from one place to another, as well as the more mundane movement of the body from indoors to outdoors, or from one room to another or even from the couch to the dinner table. The cell phone pervades the distinction between these spaces – the spatial difference between indoors and outdoors, the spatial experience of being on the street and being in a vehicle, and between familiar and unfamiliar spaces. It is also immersive, taking the body in, holding it during its connection, and opening up an exteriority or an exit from where the body is. Norman Klein (1992) suggests that when people talk on their cellular while walking, they move through the space but are not actually there. They are just immersed in their cellular; a body with a cellular, dealing with it actively, is no more in the place where her/his body is, thus the spatial differences have very little influence on her/his movement – immersion, imaginational movement.

The cell phone generates a hybrid space that merges physical and digital spaces, and immerses the body by excluding its surroundings.[23] Adriana de Souza e Silva writes, 'most of all immersion depends on imagination' (2006: 30). When someone speaks to another via her/his cell phone, s/he plunges into the hybrid space of cellular telephony where imaginational movement, exit, and exteriority are possible. This hybrid space corresponds to the potential space discussed earlier, broaching the border between reality and fantasy where experiments and discoveries may take place. In this way the cell phone echoes other transitional phenomena, rooted in 'illusionary experience' (see above, Winnicott, 1971).

As Asuman Suner (2001) notes, when we recognize that the dream of migration, movement, and departure from Turkey's landscape has become a collective yearning among its youth, the cell phone's popularity despite many of its users' low income levels becomes meaningful. From this it is possible to argue for the resonance that the hybrid space cellular telephony engenders, and to consider it as a potential space where the body's safe dwelling becomes possible, and where the body's imaginative movement and experience of exteriority become integral parts of its technological play. When inner and outer worlds are negotiated, when physical and digital space merge, the body senses this movement. It enters the space of connections – to others, and to various different spaces. An entry on the website Agora – a sort of blogger site of *Hürriyet* – confirms Suner's suggestion about cellular telephony and the desire for movement. A blogger, presumably living in the south-east part of Turkey, writes that those who come from the western regions of the country to the southeast become addicted to communication technologies, to cellular phones and internet cafes where and through which they can feel the movement to the west, a departure from where their bodies are.[24]

The cell phone also promises to materialize a personal space in which the body can feel, sense and imagine her/his individuality, maturity, and independence from familial and other authorities. The cell phone opens onto the materialization of the personal world in the imaginary. By containing elements of and about the individual, offering the possibility of connecting to others, and becoming the tool that functions for personal purposes, wishes and desires, the cell phone gives the sense that one's personal world is contained in it. Its concrete being and its material functioning materialize the dream, the imagination of privacy and individuality. The cell phone can fulfil the desire for maturity and independence or disconnection from the familial authorities, while not breaking an attachment to them. Hooke writes that using a cell phone has replaced smoking as a sign of maturity (2000). Yet it is not merely a sign or token of maturity, it is also the materialization (illusory to a significant extent) of independence and maturity, providing simultaneous distance from and closeness to others. Although the cell phone functions mostly to remove the private and personal from real life activities (see chapter six), it promises a sense of 'personal space' in which the feeling of 'individuality' becomes real. Imagining a personal world often includes the desire to have 'something' entirely one's own. I asked all my interviewees to draw an analogy between their cell phone and another machine, and a middle-aged wedding-dress tailor, Nezaket, told me:

> When I think of my life, I realize that I have always worked, ever since I have known myself. I did this and that, but always worked. Before I got married, I was taking care of my mother and some of my sisters. After I got married, while continuing to take care of my mother's life expenses, I have tried to feed my own family. There were bad times when we were about to starve, and there were good times. Now we have a car, a house…some things that I couldn't imagine having years before. But I mean I don't have anything of my own. When I think of my life, I see that I have worked to get things for others, never for myself. The only thing that solely belongs to me is the washing machine, which I generally use for the good of my family. I remember the day I bought my first washing machine, I was so happy.

Just like the washing machine, Nezaket's cellular is her personal technology. The cell phone differs, however, in that it not only functions as a personal object, but also opens up a personal space that is then negotiated alongside social space. In Turkey, young people do not often leave their family's house until they get married, or need to move to another city for employment or education. Youth rarely move to an apartment of their own when their family live in the same city. Thus it is typical to move from a place that is not one's own to another shared space. The cell phone as a personal space or inner world becomes a sort of 'life saver' for those seeking personal shelter, somewhere to dwell with one's own secrets and private matters. For instance, Nilüfer tells me:

> At the time I got my first cell, I was acting in a theatre and my family had no idea about it, 'cause they didn't want me to be an actor...Plus, because of my sexual preferences, there were people I wanted to connect with, but it was impossible for them to call me up from a home phone...I was living with my family at the time...so the cell phone has been a sort of life-saver for me... With it I was able to connect with my family and also with people I wanted to contact ...so it became something like my private, personal world.

Thus this personal space is a site in which the means of privacy can be contained and personal memories are preserved along with the potential to experience new selves and new relations. In this sense the cell phone promises to bring forth what was not possible before, particularly a refuge where the body can feel freedom. In our second talk in a group conversation on 2 August 2007 in Istanbul, Gaye said:

> I am super-comfortable with texting. I can express my feelings more freely, comfortably...The person that I am interacting with becomes a sort of 'image' for me when I am texting to her/him, so I become more comfortable, I become able to say what I want to say without the other interrupting my words, mediating my speech, you know...With texting I am like, sitting in my space, or staying in my space but also communicating with the other. The other person does not take me into hers or his. I am freer that way, at least I feel so.

Freedom here is not about movement, but rather the ability to stay in touch with others by maintaining a certain distance, by protecting personal space so that one feels neither lonely nor like s/he is invading another. The cell phone can bring those at a distance into closer range, just as it can also distance others while not breaking relations with them. Ronell insightfully writes, 'Being on the telephone will come to mean that contact is never constant nor is the break clean' (20). This is the core pleasure of cellular telephony, its promise to materialize personal space. The other in cellular telephonic interaction is a bodiless entity, whose presence is conditional for the self in her/his play space playing *fort/da* with her/his transitional object.

Cell phone relations come with a form of ethics for interpersonal encounters. One is expected to answer all calls, one is expected to reply to or offer excuses for missed calls. When calling, one is expected to display her/his identity or phone number on the screen; disguised numbers are not welcome. There is interpersonal pressure on both the receiver and caller, as instrumental to their play. When the phone does not ring, users can feel disconnected, anxious and even depressed. In a newspaper interview with cell phones users, one woman says, 'the cell phone makes me depressed at times. My cellular has to ring every day, I have to be called, asked how I am doing. These days my phone does not ring too much, and I feel so anxious and lonely' (Tu€ba ‹¾ler, *Radikal Cumartesi*, 2007). I heard similar expressions of anxiety in my interviews. Some said to me that when they are not called or sent messages they check their telephones to see if it is working properly and within range. A few told me that sometimes when there is no call and no message coming, they start scanning their address book to ring somebody, to make sure that they are still connected.

Teenagers have even developed habits for connecting: one just rings the other and in return expects to be rung back. Without talking or texting, just by ringing ('beeping' says Donner, 2005b) they say to each other 'I think of you,' or 'Is everything okay?'. In essence this shows that they are present in the other's network, a simple message of 'Do not forget me'. Older users have a similar practice of calling the other, speaking very briefly, asking how the other is doing – no instrumental reason other than making sure others are still there, still attuned, still in the same network. Couples use their cell phones to monitor their partners, constantly checking where they are and what they are doing. Parent-children relationships follow this same pattern. As a part of daily coordination and arrangements, the self and others come together to ensure that the connection is there, that the relationship is active.

The cell phone's instrumentality rests in its ability to assuage the anxiety that others are there within the play, still serving the dream of the process of individuation and maturation, helping to build a personal space which is both connected to the social world and separate from others. Translation in cellular telephony can be approached through Bruno Latour's conception of translation – the artefact encourages certain actions while discouraging others. The caller tends to keep the interaction short. The mini-keyboard, mini-screen, low memory space and high taxes on extended use all discourage lengthy interactions with others.[25] By encouraging brevity – short conversations, messages, videos – the cell phone mediates the relation between self and other in a way that prompts the caller and the called to further and physical interaction, for stronger, more intimate relations. Cellular telephonic contact simply reassures the pair that both bodies are there in the play. As Christian Licoppe's research shows, the SMS messages that circulate between individuals are most often forms of 'I'm thinking about you', 'I am with you', and 'What are you doing?'. My observations of the Turkish mobile technoscape confirm this tendency. When considering these interactions in terms of their observable emotional contents, one may think that they say nothing. However, as Callon and Law (2004) have suggested, by virtue of their circulation they are also saying everything. The frequency of calls, the continuous circulation of these messages perhaps create love. The self and the other in cellular telephonic interaction assure each other that the play is still going on, that bodies at both ends are still in the play space, that each can accomplish their own project.[26] For this reason communication needs to occur at the border, where distances are (ideally) more easily protected. Play with the cell phone becomes a safe play with others, for the other is a bodiless entity, an image that fosters possibilities and allows one to express what one wants to say.

Cellular telephonic relations between self and other are always pregnant with possibilities; they work by interrupting the body in its usual movements. When the cell phone calls, rings, or vibrates, it actually interrupts the body's rhythm and movement, action is put on hold as the cellular starts its material functioning. This interruption promises change: a change of plans, adding plans when there are none, a shift in affective situation, or a movement from where the body is. The cell phone's performance in everyday life offers an escape from long-held plans, and also from punctuality, stiffness, and certainty; it provides the sense that life has room for spontaneity, flexibility, and improvisation in interpersonal and intrapsychic relations.[27] GSM providers and cell phone manufacturers often emphasize 'living the moment' or 'seizing the moment' in their advertising campaigns. The moment they promote is the moment of possibilities, which at its most utopian works against the grain of norms, certainties, pre-planned actions to provide the sense that one is about to experience what one can otherwise only imagine. In a group conversation, Kaya and Nalan both agreed that the cell phone means possibility for them. Kaya told me that a call can change all his plans, or add some plans when he has none. This, perhaps, prompted Nalan[28] to say, 'perhaps, we are attached to it, despite of [sic] everything it does to us, because of possibilities it might bring. Sometimes you just wait for a call, a sort of unknown call that would make you feel good.'

By interrupting the body and opening a potential space to negotiate inner and outer worlds, by offering a means of improvising with others, of translating one's own language and the other's simultaneously, cellular telephony promises new ways of experiencing the old psychological/emotional contents. It is instrumental for psychic individuation. Cellular telephony enables the self's projective identification by implying possibilities. It provides a space for the endless play of *fort/da*, and to store 'pieces' of this play. In its containment, pieces of self, memories of lived experiences (photos, messages, videos) and the possibilities for the not-yet-lived come together to form an articulation with which one attempts to identify.[29] We should recall Donna Haraway's (1992) contention that articulation is a work in progress that contains the possibility of failure.[30] Failure is intrinsic to the work of the technology that I have explored here; because the arrival to self-proper is impossible, working through the cell phone towards an imagined articulation or individuation as a subject (self-present) is always bound to fail. Yet as I argue in chapter three, this failure and its repetition is the very condition of the work.

As a technology of transference, translation and containment the cell phone necessarily speaks to the search for life experiences through which one can feel at home. Like the telephone, it evokes a sense of shelter, where one can be contained, where life is preserved. If, as Ronell suggests, the telephone's transference and translation swallow what is not there, the cell phone is a sort of wearable and miniaturized form of this containment. It not only houses possible experiences, holding the repetitive movement of *fort/da* for those who feel out-of-space, but it also contains the traces of one's lived experiences, memories and the stuff that one needs to see and feel an (illusionary) sense of progress towards the imagined arrival. This is one of the cellular telephony's performances that makes this technology distinctive for the play of *fort/da* – it contains the 'selected' metaphoric images, messages, videos, and sounds that belong to the 'person' and that suggest to the body that 'self-reflection' is possible. Its distinctive promise is that it does *fort/da*'s work for the melancholic body, at any time and in almost any place. It provides a sense of life-giving movement to bodies whose ears are trained by the discourses of 'stasis'. It gives a sense of connection to others, which is required to be a living-body in the social world, especially where psychic connections are politically, culturally and economically precluded. It is highly addictive.

The way cellular telephony moves us in Turkey is strongly related to our collective will to live multiple lives, to borrow Helene Cixous, (1986: 112) description of identification. We are so tired of only being able to come to the social space by destroying personal spaces; we are so tired of being bound to descriptions of ourselves within the limited and limiting social categories of *kıros*, non-modern Islamists, grown-up adolescents, immature young ones, 'liberated' secularists, unconnected (incommunicable) Kurds. In Turkey the cell phone is our toy that unfolds a potential space where the personal is ideally protected and yet the social bond is still felt. The cell phone is the toy that promises to interrupt our collective melancholic condition by urging us to sense the possibilities of its containing functions. The cell phone is the toy-like vehicle which instigates affective and imaginative movement; it takes us into its hybrid space, which corresponds to the potential space for our very articulation. It speaks to the 'third world' children[31] in all of us by promising movement, maturation, individuation and connection.

6

CELLULAR TELEPHONY AS A SOCIAL PRACTICE: THE COLLECTIVE DESIRE FOR LIVING IN AN OPEN CROWD

> You know how we used to define ourselves, with horse, woman, weapon. I think for contemporary Turks, the cellular took up the role of weapon. This [the cell phone] is a weapon for making crowds...not a simple weapon...when you have your cellular, you're part of something, like a crowd,[1] but you're also immersed in your own thing (Ada).[2]

Elias Canetti (1984) opens his book on power and the crowd by saying, 'there is nothing that man fears more than the touch of the unknown. He wants to see what is reaching towards him, and to be able to recognize or at least classify it' (15). For Canetti, this fear of touch is at the foundation of societies in which fear distances bodies and ranks them in a certain hierarchy – which in return generates more and more distance between bodies. Distaste for the other's touch is the origin of hierarchies, rules, and status within society. Moreover, Canetti argues, the fear of touch is also the reason for crowds to gather. He writes, 'it is only in a crowd, [that] man can become free of this fear of being touched' (15). In the crowd, the fear changes into its opposite, into the pleasure of being together with other bodies, where (ideally) all differences, including sex and race, dissipate into the density and movement of the crowd. Canetti, a 'master' (Bogard, 2000) at theorizing crowds, is particularly concerned with the physical crowd, but the conscious and unconscious mechanisms that he uncovers also determine the invisible crowds of people that share a certain dream which gathers them together – unconscious, unplanned. For Canetti, the dream of the dissolution of all differences creates the crowd. He writes:

> The important occurrence with the crowd is the discharge. Before this the crowd does not actually exist; it is the discharge, which creates it. This is the moment when all belong to the crowd get rid of their differences and feel equal…it is for the sake of this blessed moment, when no one is greater or better than another, that people become a crowd. (1960: 19)

Man temporarily loses the fear of touch, gives up his identity and becomes part of a larger force outside of his control, because the density of the crowd is capable of releasing the burden of distance where body is pressed against body. Crowds are always made of multiple individuals, in fluid and changeable configurations. Individual differences are not important in the movement of the crowd; it only matters what circulates between and through individual bodies, the flow of heat and energy. The crowd acts like one body or one self, like a flock of birds whose simultaneous flying movements make them seem like one thing in the sky.

Ada told me that the cell phone is a weapon for gathering crowds in Turkey. He used the word 'weapon' to define the cell phone, which gathers crowds in Turkey for another reason which will become clearer below. He did not refer to physical crowds of people coming together in a particular space in the form of 'flash mobs' (Savage, 2003) or 'smart mobs' (Reingold, 2002) to express a political collective agency or to demonstrate against particular political or social state policies. Even when we witness 'smart mobs' in Turkey's political sphere, cellular telephony is not declared as the main factor in generating these rare crowds. Some studies claim that cellular telephony is extremely useful for creating physical crowds in which different individuals come together to share the same dream of action and to express it collectively. For instance, Vicente Rafael (2003) talks about the crowd of people who took to the streets of Manila with a will to bring justice and equality to the Philippines' social sphere. He views the crowd and cellular telephony as two technologies which speak to each other, which serve the same purpose of conveying a collective political view. Cell phone politics have also been associated with 'smart mobs', 'flash mobs' and other physical crowds where the cell phone takes on the role of amassing a number of different people to express a collective will, desire, or dream (Plant, 2000; Rheingold, 2002; Savage, 2003). An ubiquitous personal object that the majority of the population owns, the cell phone is well-suited to the mass circulation of decentralized messages, calling people to collective action by joining a crowd. Yet I did not come across any stories in either my search of the media or my interviews with cell phone users that would indicate cellular telephony is actively associated with political or social crowds in Turkey, as it is said to be in the Philippines or in Zimbabwe (see chapter one; see also Rafael, 2003; Pertiera, 2005; Sheller, 2004; Agar, 200).

Ada was not referring to the virtual gathering of people in the mode of a 'juridical crowd' (see Gittlin, 1986), where different people form a social body by sending SMS messages to become part of television shows, for instance by voting for the elimination of one of the competitors from series like *Big Brother, Pop Star* or *Turkish Alaturka3 Icon*. Nor did Ada mean to speak of the crowd as an audience comprised of individual bodies that, as subscribers to GSM companies, become part of the whole addressed by decentralized messages. One can receive various text messages throughout the day from an unidentifiable sender, and one cannot return the message. In this way, cellular telephony has distinctly commercial ends through the circulation of such ads, a communicative potential that can also be used for parties' political propaganda.[4] Ada spoke of the crowd where people feel connected and yet are able to enjoy their solitude, where they can be immersed in their own world while still being part of something larger than themselves. He gave an example:

> When you are on a *dolmuş*,[5] you are with all those people, squeezed in the small place, pressing on each other. And almost all these people take out their cell phones from their pocket and start playing with them, calling somebody, or texting…So when you are on the *dolmuş* all together, there is nothing that you share, even you feel disturbed by others' presence. Smelling the sweat, feeling others' arms, bodies that close to you…but when you play with your cellular, nobody speaks to you, there is no interaction between people, so it is like a weapon for putting a sort of distance around you, protecting oneself from others. But on the other hand, because everybody is doing something with their cell phones, you're alone but also part of something…it makes a perfect crowd.

Civan,[6] an interviewee who owns a gas station in rural Ankara, reflected on another aspect of the crowd that is produced through possessing cell phone and through engaging with cellular telephony as a social practice. His observations can shed some light on our discussion of cellular telephony and sociality in Turkey:

> When people have a cellular like others have or the same machine as others they feel equal, as if they become the same. Especially those poor, people from *varoş* [outskirts of cities], when they get a cellular and use it for the same purposes that for example I do, they get the sense of equality. I think the main reason for it to become so popular in Turkey, is that we as a nation feel so inferior, so when we have something in our lives that we think, for instance, westerners do or upper classes have, then we get this feeling, that yes, we are equal, we are the same. It is like a sort of revenge...

In chapter four, where I discussed how the meanings attributed to cellular telephony and its performances are as a means of social differentiation, I maintained that cellular telephony also works to establish commonality – that the massive engagement with cellular telephony by collectives that define themselves or are defined by others in terms of their differences also manages to display a sort of communal identity. In this concluding chapter I discuss commonality in terms of how we can understand its production through cellular telephony, how we can identify forms of sociality within this commonality, and so comment on the collective desires, imaginations, inclinations, responses, and affectivities that are wrapped up in the promise of a certain sociality or certain social relations of the cell phone's performance.

I opened this chapter with a discussion of the crowd to pose the question of whether or not we can see the sociality evoked by cellular telephony as a crowd, whose motive is the dream to discharge differences so that all bodies can feel a sublime equality through the flow of energy between these bodies in the density of the crowd. Yet this provokes further questions. Even if we assume that as a social practice cellular telephony encapsulates a collective desire to feel such moments through an engagement with the performance of cellular telephony, to what extent can we argue that it actualizes Canetti's dream of generating crowds? If we assume that cellular telephony promises to create a crowd in which bodily relations are not bound by strict identities that conform to the given social categories, how is this sense of potentiality accomplished? Such questions cannot be answered without discussing the spatiality of cellular telephony.

The issue of sociality needs to be examined by taking into account the space where the means of sociality takes place. Spatial characteristics determine the context of sociality, and mediate the means of social relations. In other words, we need to investigate the kinds of spaces bodies inhabit as they engage with cellular telephony and as they become part of the sociality it produces. In the previous chapter, I highlighted authors who claimed that cellular telephony opens up a hybrid space where the physical and the digital interact and interfere with each other. I argued that this hybrid space, which works through immersion, contributes to the potential space for the individual body where s/he safely plays her/his own game of *fort/da*. With this in mind, I seek to explore the ways in which this potential space becomes a collective one, where individual bodies imagine, sense and experience the means of sociality. Two models of spatiality, namely the network model and the fluid model, will be my main foci in the following discussion, as I examine how and why the cell phone's social promise and work encourage attachable technological performance. Spatiality, as what Law and Bijker (1992) term a socio-technical 'geography of enablement and constraint', cultivates some forms of structuration over others. While network spatiality means rigidity, stabilization, identity and control, fluid spatiality can be considered as that which favours 'a way of letting go, instead of holding onto, the rigidities of network' (Law and Mol, 2001: 653).

In the current literature on cellular telephony, the space that enables and constrains contact, touch, and co-presence between physically distant bodies is often understood by a network metaphor. Within this model, bodies are seen as stationary nodes and their embeddedness in space as a condition for their identity construction (see Castells, 1996; Lash and Urry, 1994). One connects with others, or more precisely, one experiences the means of connection insofar as all the entities that make up the network are actively embedded therein (see Sheller, 2004; Urry and Sheller, 2006). The network model highlights the order and control hidden in the mode of flexibility whereby bodies become modulations of coded information that flow through and become linked to each other (see Bogard, 2004). The network records and tracks this movement, and regulates the relations between nodes. Not only does it track this flow, but the network can also exclude individuals, groups, collectives, and even nations, controlling and stabilizing global and local social structures (see Castells, 2000). As a framework for approaching the means of order, control, and regulation that are disguised in the flexible modules that enable the movement of information, data, images, bodies and capital, this model has been very useful and valuable. As a social practice, however, cellular telephony in Turkey draws attention to significant limitations and interruptions of the network model. Cellular telephony de- and re-territorializes the network by being a much more fluid technology than the network model allows, that travels through fluid spaces and so evokes fluid socialities.

According to the network model, cellular telephony produces social relations between different bodies, enables the flow of information through them by stabilizing their identities, controlling and regulating their movements. From this perspective, cellular telephony is a means of 'identity and control' (White, 1992; Law and Mol, 2001; Sheller, 2004), that anchors relations between bodies. Moreover, as a global networking technology, cellular telephony seems to be part of what controls and orders connections between the global and the local. If we understand cellular telephony as something that works only through networking, that remains the same as it traverses a global network, then the cell phone appears to be a standard product without provisions for variants in use, meanings, or shape.

As I argued in chapter three, a tool does not become a social practice simply through the standard features of its technology. It becomes a social practice insofar as it speaks to collective and individual desires, imaginations, inclinations, responses, and affectivities. Especially in countries like Turkey which import and domesticate technologies, the tool's capacity to adjust to the local technoscape determines its degree of domestication. I maintained in chapter one that when technology, which is an invention and emerges from particular social, cultural, and economic conditions, is imitated – meaning that it becomes a social practice – it becomes different but also retains similarities to an invention. Imitation implies a unity of reproduction and difference. A Nokia machine designed and produced in Finland is the same machine that travels to Turkey, yet it also changes as it becomes imitated. That is to say, the way it is used, the meanings that are attributed to it, and the task that it performs are necessarily different in different cultural contexts.

A network-object is assumed to travel through the net by holding its shape; therefore the object that travels the network is categorically the same object (see Latour, 1987).[7] For instance, as a network-object the cell phone travels through the global network and because its shape does not change, it is still the object that communicates with other cell phones in the network. However, Mol and Law (2001) write that as the object travels through the network, not only do its use and performance change but its shape as well. In line with Mol and Law (2001), I suggest that the cell phone's shape changes when it is domesticated. Pieces of the standard machinery are altered – plugs are different, software programs are different – so that it speaks the local language,[8] some machines are programmed to ring at prayer times, and the mobile systems that run machines change (Turkey uses the European GSM system). Features change as well: possibilities for accessorizing (colourful covers or stickers), ring tones (theme songs from local television dramas are often used as ring tones in Turkey), and even the inside of the machine is altered as its containing space contains different things (see chapter five). Thus the Nokia machine used and possessed by Turkish bodies in the Turkish technoscape may look like the Nokia machine that the Finnish, British, German, and Japanese use, but it speaks a different language, it has different configurations, its plugs are different, its insides, its meanings, its performances, and its tasks are crucially different. All cell phones are portable, personal, pedestrian, and all of them are designed to make human involvement in its machinery – to a certain and variable extent – possible. All have keyboards, screens, and software programs that work with particular infrastructure, and demand the hand's work. These standards let us categorize it as a cell phone. Yet as it becomes domesticated, imitated, collectivized, and personalized, it changes. One of my interviewees, a bank officer, said when he showed me his Nokia:

> Some Finnish guy produced it, it came through all the way to the territory of Turkey and my wife, my brother and I use it. I regret that we, as Turks, couldn't invent such a thing. But what we did is to adopt it to our culture. Family is important for us, so we

> constantly speak with our family members, even those that you don't see too often…Or *bayram* religious aid is important for us, we send celebratory SMS to everybody in our telephone book. Or for worse, we use it as something that shows others how clever we are, how modern we are, how rich we are…Or you know this minister[9] who falls asleep in all meetings, for instance, his friends call him to wake him up…We're good at finding some strange ways of use for these technologies, but not good at inventing a technology…we lack the self-confidence, this is the whole issue.' (Sadun)

As a global technology that is produced and invented somewhere else, cellular technology is a foreign technology. When cellular telephony is recognized as part of the global network whose felt and perceived centre is not Turkey, it continues to foster historical resentment, anger, or melancholia. Sadun says, 'I regret that we couldn't invent such a thing' and the reason is that 'we lack the self-confidence'. Cellular telephony shows what 'we lack' insofar as it allows for varieties in imitation, in use, in shape, in its performance.[10] Although most of my interviewees, in line with many journalists and columnists, depict the way the people of Turkey engage with cellular telephony as something that shows an historical 'inferiority complex' or 'ignorance' (see chapter three), at times the imagined Turkish stamp on global cellular telephony represents what differentiates Turkish cellular telephony from others as an ability to creatively adapt the technology. One humorous television advertisement for Akbank banks is an exemplary reflection of how cellular telephony is envisioned as marking an imagined difference between the local and the global.

On a fast train,[11] an American, a Japanese, and a Turkish guy sit at a table, having their lunch. The American with a cowboy hat shows off his cell phone and expresses his pride at having invented this technology. The Japanese fellow, who has a flying cellular,[12] says that he is proud of developing and improving this technology. The Turkish guy, who wants to end the discussion with his display, says that he will pay for their lunch – he removes his cellular from his pocket, contacts his bank, takes out a loan (apparently, a lunch on that Turkish fast train is prohibitively expensive) and pays the bill immediately. He declares that he is proud of inventing new ways to efficiently use this technology. In the end, both his American and the Japanese friends accept that the Turkish invention is the greatest. The Turkish guy recommends that in their own countries they work harder to catch up with Turkish inventions.

We can read this ad as a result of an unsuccessful advertising strategy which could not predict its own reception, as the ad implies that the stamp of Turkish difference would appear when it came to finding new ways to take bank loans. Yet even as a poor ad, it represents core aspects of my previous discussions of the technoscape, melancholia, and the ways in which cellular telephony is seen as a means to overcome the historical melancholia in the social imaginary. This ad illustrates how as a contemporary performance of the historical dream of presence, cellular telephony opens up a site where the gap between the local and the global is imagined and accommodated. Historical melancholia and the historically imagined western gaze pervade the commercial, contributing to the popular perception that cellular telephony can only become a means of national pride when it garners the approval of the imagined western gaze – in this case the imagined global gaze of the global community. Locality becomes meaningful insofar as cellular telephony can link Turkey to the global network in ways that allow for Turkish national pride to enter the sphere of globality,[13] a vision that we have long dreamt of being realized.

The global and local negotiate with each other in cellular telephony's hybrid space, where it both participates in and against globalization (see chapter three). Cellular telephony in Turkey is a contemporary performance of the technoscape in which the imagined western gaze has diffused into the imagination of globality. Since cellular telephony is global, it is worth localizing, and because it is global, it can be the very site where the gap between the global and local can be faced, and where an imagined 'Turkishness' can be played out. If cellular telephony refers to a global network whose centre is far away, then its localization and domestication – the way it is used in the landscape of Turkey – can mark and produce difference, a national pride.

To me, the ad described above tells more than it initially meant to express. The Turkish inventor who sits in the 'centre' of the global technological network, on a high-speed train (which remains a failed project), exhibits his 'difference' through an invention borne of imitation, and becomes an object of admiration to the global community – ironically, through inventing a new way to apply for bank loans. Given that Turkey's economy is dependent upon external and internal debts that are administered according to the IMF's agenda which forces countries to fall further into debt, this irony deciphers the truth of globality, that the means of movement and mobility are not available to all. This global technology territorializes the network (through the global and technological capital of its centre), and spatializes the centre and periphery (some produce and improve the technology, others consume it), yet when domesticated it also carries the potential for deterritorialization. It engenders possibilities for transformation through imitation and fluid configuration – varieties in use, imitation, imagination, performance, and shape – which hold out the possibility to change how the global network is structured.

In chapter three, I have commented on the ways in which an engagement with the global technoscape determines a sense of the landscape, and that this can unfold as collective resentment or collective pride. I have also suggested that the technoscape provides a means for marginalized countries, collectives or individuals to demand their share from the globality. Also in chapter three, I have written that the cell phone has been more passionately and quickly adopted by 'developing countries' than by developed countries. I have argued that Turkey's technoscape needs to be examined as a historical dream of presence in which melancholia can be overcome. Here, I will add that cellular telephony is not merely a network technology, linking the local to the global, but also a fluid technology, allowing for heterogeneous imitation and use that can become a means to assert national difference. Cellular telephony is an appealing, passionately adoptable and even collectively attachable technology in marginalized countries because of this fluidity, which sanctions the ways collectives can become 'producers' of technology through their own experiences with it (see chapter four).

Cellular telephony thus exceeds the network model. As many authors maintain (see Chesher, 2007; Sheller, 2004; Donner, 2005a), fluid models directly address the complexity of cellular telephony, including how it changes from one place to another and takes on different roles, meanings and tasks. However, its complexity is not fully encompassed by the fluid model either. Rather, the spatiality of cellular telephony needs to be understood as a hybrid space in which both network and fluid forces circulate.[14]

Like a network, fluidity is another metaphor that social scientists have proposed for examining the contemporary mobility of technologies, things, capital, bodies, and sociality. Authors working with these issues claim that the means of mobility and sociality are more complex when relations between nodes are unclear or uninterrupted, and that nodal unities are not simple to delineate. For instance, Nigel Thrift, a leading scholar in geography, writes that in classical network theories, technologies are assumed to be seamless systems 'without interruptions and limits, presented as coherent and consistent, without difference on locality'; they are assumed to spread anywhere easily and quickly because they are 'mainly comprised of representation, rather than technical repertoires in use' (1996: 1468-69). In using this chapter to extend my previous discussions of how cellular telephony moves and gathers people, I will reiterate that cellular telephony is not solely comprised of representations, associations, signs and meanings, nor is its addictive performance constituted by its instrumental functions.

Cellular telephony is a fluid performance that adjusts to the conditions I have described through the concepts of technoscape and melancholia. Not only is it fluid in terms of enabling affective and effective domestication, but it is also fluid in how it lets individual bodies experience, sense, imagine, translate, transfer and inhabit their singularities in its containing space, which is key to the project of arriving at the self-proper through improvisation with others. I continue to insist that as a 'technical repertoire', cellular telephony requires human involvement in its machinery, at the same time as it asks the human to be the 'producer' of her/his technology/drug/commodity. Cellular telephony is both fluid, travelling through fluid spaces and generating fluid social relations, as well as a network, generating the means of 'identity and control'. John Law and Annemarie Mol explain, 'fluid':

> The fluid metaphor suggests that we are dealing with something that is viscous: with things that tend to stick together. But it also points to a possible difference – a difference between fluid and network spaces. For in a network things that go together depend on one another. If you take one away, the consequences are likely to be disastrous. But in a fluid it isn't like that because there is no 'obligatory point of passage', no place past which everything else has to fire; no panopticon; no centre of translation; which means every individual element may be superfluous. (1994: 661)

In the space of fluids, bodies and machines can be performed in more than one way. Accordingly, as a fluid object travelling through fluid spaces, the cell phone will engender social relations in which bodies interact in complex ways that cannot be understood as clean and stable ties. In fluid spatiality, bodies – their identities and movements – entities and machines adapt or adjust to the conditions in which they travel.

Cellular telephony's hybrid space contributes to the possibilities for social relations that are created when both network and fluidity negotiate and circulate with each other. When sociality is described according to the network model, each coupling, or cello-technological touch in the net restates the means of identity and control; the fluid model depicts the potential to put given social identities on hold, to deform strategies of control, and to generate new social structures (Sheller, 2004). Mimi Sheller's (2004) fluid model examines socialities enabled by mobile communication technologies. She argues that mobile communication technologies have merged the means of transportation and communication, and enabled bodies to connect when they are on the move. By reviewing Harrison White's model of social gel, Sheller writes that in mobile publics, sociality follows the logic of a sort of messy gel where bodies get in touch with each other, couple and decouple constantly and continuously, and so can slip from one identity to another. She maintains:

> I want to suggest that the increasingly mobile conditions for communication are enabling a new kind of public-private, a kind of fluid social space in which communication occurs which spans absence and presence, personal and impersonal, micro and macro, local and global. (2004: 47)

Sheller's notion of 'coupling and decoupling' emphasizes the latter, for she pays close attention to when the network ties are loosened or decoupled from social contexts as well as from social spaces and structures. Her model is overly optimistic and utopian at times; for instance, she writes that mobile communication technologies produce mobile publics 'hitherto unimagined, unpredictable', that can challenge traditional notions of structure, social space and scales. As I explained earlier in this chapter, cellular telephony cannot realistically be understood as something that produces mobile publics where the social structure changes in an unpredictable or unimaginable way. The spatiality of cellular telephony is not and cannot be fully free of network spaces, where the control and stabilization of existing social structures prevail. It is precisely the interaction between network and fluid spaces that make the cell phone an attachable technological performance in Turkey.

Rather than suggesting that cellular telephony produces a form of sociality where the distances (and their concomitant hierarchies) that preclude coupling or touching are overcome through coupling and decoupling, I argue that cellular telephony's hybrid space is where society is reproduced and a crowd is potentially produced. This opens up a site for social struggles wherein networks are collectively re- and de-territorialized. Before elaborating further on this, I would like to reflect on the ways in which Turkey's cellular telephony market has developed, and the kinds of marketing strategies GSM networks have adopted throughout this development. Examining this context of cellular telephony is important, as marketing – including pricing, a network's penetration capacity, communication strategies, and changes in GSM services – determines to a significant extent how cellular telephony functions (e.g. produces distinct networks or fluidity across networks) as a social medium in social space.

In the first five years of Turkey's cellular telephony market (i.e., until 2001) two Turkish GSM providers, Turkcell and Telsim, enjoyed the highest profits in this growth market. Turkcell and Telsim have so effectively spatialized the social space that each network has become associated with particular social and economic classes, which can themselves be labelled as networks. While Turkcell has been associated with Turkey's wealthy, urban elite, Telsim connotes the people, periphery, and 'third worlds' of Turkey. Their ads deliberately reflect and reproduce these associations.[15] Telsim is less costly yet has a low penetration rate, while Turkcell is more expensive with a higher penetration rate. Both companies are more effective at enabling smooth contacts or couplings via cell phones in urban areas. Additionally, both GSM providers to charge higher fees for inter-network contacts.[16] Thus the marketing strategy of these two networks is to create distinct social spheres across which interaction becomes costly.

Some of my interviewees told me that Turkcell has always been more prestigious than Telsim because it promotes itself as being the modern face of a Turkey that embraces the future. Some also mentioned how having an early Turkcell number functions as a source of cultural and economic capital, and that it can affect relations with people or even institutions. For instance, Tarık[17] recounted that 'the Turkcell numbers that begin with two are on the market for $10,000; because if you have a number like that it means that you are one of those who purchased cellular telephony long time ago'. Civan attests that 'banks officers look at the cellular numbers of people who apply for bank loans, and ones who have Turkcell numbers that start with two or three gets the loan almost for sure'. Emine, who has a prepaid Avea line that she uses to interact with her family and friends, in addition to her Turkcell line,[18] says, 'when you give your prepaid line number, people may think that your business is not good enough to let you afford a post-paid line'. Although I have found no proof (with some exceptions)[19] to support Tarık's and Civan's claims[20] that bank staff indeed pay attention to telephone numbers or that numbers that start with two have been sold for thousand of dollars in the illegal market, their statements – including Emine's – show that these particular GSM networks have strong and particular social, cultural and economic associations in Turkey's social space. However, as new networks have been introduced,[21] the market has become more competitive and consequently the fees for intra-network and inter-network telephonic interaction and SMS contacts have decreased significantly. The marketing context has changed, extending the means of mobility in that sociality across different networks has become much more affordable.[22] Additionally, in the new mobile communications market, the previous distinctions between the meanings attributed to GSM networks are not as clear as they were in Turkey's early period of cellular telephony.[23] Hence, the contemporary marketing context of cellular telephony allows for increased fluidity, particularly by easing the coupling across GSM (and therefore social) networks.

Against this background, the ways in which cellular telephony becomes a means of both re- and de-territorializing networks in social relations becomes significant. A personal object that can be carried anywhere, the cell phone is a perfect instrument for familial, corporate and governmental control, as it can track and even regulate bodies' movements, and interactions. Parents control and survey their children, 'remote mothering' (Geser, 2003), via the cell phone. State organizations, GSM companies and corporations track subscribers' interactions and movements through their cell phones, and in a sense this work can be aligned with that of the panopticon. A body with a cell phone is always subject to the smoothing marks of the social and political power, as William Bogard (2000) argues. Once you have a cell phone, you are expected to take all calls. Sometimes, refusing to take an unexpected call is considered an act of irresponsibility. As Süreyya[24] noted in a focus group discussion with her husband and Emine (Süreyya is Emine's boss):

> If you have a cell phone, then, it means you are always available and accessible to others' calls. To answer a call or return the call by giving an excuse if you couldn't reply is one's responsibility. It should not be understood as a freedom to reject others' calls, it is a freedom that makes people contactable, communicable when they are distant.

Cell phone conversations often open with the now-standard question 'where are you?' (see also Licoppe, 2004), which is often answered by the other person giving her/his coordinates. Not only does the content of coupling or contact between people enable interpersonal control and reinforce interpersonal pressure but the frequency of coupling with the same person can become a means of interpersonal control and pressure as well, even if it is intended as a means to show one's love or care (see chapter five). For instance, most of my teenaged interviewees told me that they are fed up with their parents' constant and continuous calls. Burçak,[25] a high school boy, says:

> I don't know how many times my mum calls me on my cell during a day. She calls me and says, 'where are you, who are you with, go home and do your homework'. And I say, 'yes, mum, ok mum'. I told her many times not to call me that often, but she still calls me every day, almost every hour during a day. She says she gets anxious when I'm not home, but the truth is she tries to find out what I am doing, whom I am with.

Cellular telephony extends the means of coupling, inserts the ability for constant coupling, and thereby has the potential to function as a vehicle for stabilizing identity and exerting control. In each coupling, the means of identity (e.g. a son and a mother), of control (the mother is able to track, to a certain extent, her son's movement and can continue to give orders), and of pressure (a mother who regularly calls to give orders is not only a teaching or commanding mother but also a loving and caring one) come into play. In this regard, each coupling may be considered as a form of social power that marks the body, smoothing it to fit into the given structures of social relations. Süreyya said to Emine, 'I am older, it is a courtesy if nothing else to respond my calls', after Emine argued that she is free to not take all cell phone calls. My father often says to my younger sister that she cannot reject or put off his calls because he is her father. A commercial director, Hidayet, told me that he always takes his producer's calls, because his producer provides contracts for him.

Thus cellular telephony, as Ronell (1989) says of telephony, is an affirmation of duty, readiness for the order, acceptance of the subjection, saying yes to the name that is uttered. Taking the other's call is like entering a contract, affirming the name and identity imposed on the body: my sister is the daughter, Hidayet is the commercial director, Emine is the younger; or else Hidayet is the name of the body that answers the call or is addressed by the caller, or else Emine is the younger woman and the one who works for Süreyya. Cellular telephony thus stabilizes identities, enforces networks, controls movement – it smoothes bodies to make them fit into social categories (customers, subscribers, Telsim subscribers, prepaid line holders, *kıros*, 'third worlds', 'immature teenagers', 'Islamists' whose telephones ring when the pray times come, 'Kurds,' woman, man, et cetera).

Yet, if cellular telephony worked solely in this way it would not have become an attachable technology in Turkey. Cellular telephony has become an addictive technology because it also puts identities on hold, it opens up a site where otherwise non-existent means of social relations are potentially liveable. If on the one hand cellular telephony can be defined as the reterritorialization of the network, as I have outlined above, it can also be said to favour deterritorialization, through the potential emergence of mobile crowds. Before identifying the potentialities enabled by cellular telephony to put identity on hold or to move across multiple identities and multiple social contexts (decoupling), I will address one further aspect of coupling. Specifically, I will outline how coupling mediates the experience of co-presence, absence, presence, distance and proximity, and how this limits the enforced 'identity and control' that is arguably intrinsic to each coupling.

One couples with another via cell phone through telephonic interaction, texting, sending audio or visual messages. Thus coupling refers to the contact or touch between bodies that are otherwise physically distant. Coupling lets distant bodies touch each other. It brings the possibility of touching and being touched while remaining distant. With cellular telephony one is always accessible and available for coupling, but also for decoupling. When one couples with another, one senses presence as well as absence, one understands proximity as well as distance between them. Thus in the course of coupling, bodies are neither present nor absent and neither fully distant nor fully proximate. William Bogard (2000; see also Urry, 2000; 2004) writes that communication in cyberspace, the experience of coupling with another who is absent/present, in distance/proximity, is uncanny:

> [S]trangeness...contradiction between nearness and remoteness, or mobility and fixation...Cyberspace communications, in a word, are strange – at the push of a button, territories dissolve, oppositions of distant and close, motion and stasis, inside and out, collapse; identities are marginalized and simulated, and collectivities lose their borders. (2000: 28, cited in Urry, 2002: 267)

This strangeness intensifies with cellular telephony, for the experience of distance and proximity, absence and presence is so intense that at times it may reveal the very impossibility of 'identities', 'selves' and 'self-presence'. Papatya[26] narrates her strange experience with cellular telephony as she and her lover attempted to compensate for the loss of physical touch with a cello-technological co-presence:

> This thing has started between us through touch. I mean our very love was very much based on touching each other, seeing each other, speaking face-to-face, being present body to body. In the beginning, both of us were in Istanbul, so everything went very well. But after a while, I had to travel for my work. And the relationship became actually cellular telephonic love. I remember, once all night I talked to him on the cell. Toward the morning, I even fell in sleep with the cellular at my ear. But throughout these long and short talks on the cell, we just felt that this machine actually was killing the magic of our thing. It was not so much, not seeing and touching each other, but rather trying to compensate this absence with cell phone killed it. As we were missing each other and not seeing each other, we were calling each other more frequently. And each time, his image in my mind was transforming into a sort of voice in a dark. He became more and more bodiless person for me... I don't know if it makes sense but I guess with cell phone we [in English] 'lose our physical consciousness'...it's totally not real, it is something *yapmacık* [superficial].

The question of whether communication technologies becoming more haptic is replacing the sensuous means of touch[27] or corporeal co-presence is an old question, to which dystopian technological determinists have emphatically answered 'yes', as these technologies supplant physical contact and cause a decline in social capital (see Putnam, 2002). If we think of this question in light of Papatya's experience, we can surmise that virtual means of touch cannot replace the sensuous means of touch nor can virtual travel substitute for corporeal (see Urry 2002). Communication technologies do not necessarily replace existing means of touch, corporeal co-presence or existing social connectivity; they introduce another dimension where an experience of what is missing is partly (through imagination and hallucination) possible. Yet as we see in Papatya's statements, the uncanniness of cellular telephonic coupling can also obstruct the imagination and hallucination needed to accommodate this lack. As much as cellular telephony serves the illusionary play of *fort/da*, it may also reveal the impossibility of this play. The uncanny absence/presence, distance/proximity, intimacy/distance might well uncover the phoniness of the names, identities, or social categories. As Papatya's lover transfers into the digital sound that speaks to her, his bodily image gets lost, he becomes 'a voice in [the] dark'. Ronell has observed similar sensations about the telephone:

> The phone phones, shading in a differential register of inauthenticity, establishing the phony, the shady Other, like the moon, whose identity and therefore also ours is held in suspension. 'Hello, may I speak to?' 'You are' so the voice that comes from me and beyond me can be a phony one, it can miss the point, performing and inducing fraud, putting a metaphysics of identity on hold. (1989: 45)

Thus, the coupling of cellular telephony might put identity on hold, it may challenge subject positions, and so allow one to improvise, to play other roles, outside of what is given. This experience can let bodies escape from the terrain of social categories, given identities, or given social relations. Savaş, a teenaged high school student, told me that his parents, just like his friends' parents, are too controlling, that they do not allow him to express his individuality. He hardly manages to live with his father's aggressive authority and with his mother's excessive hypochondria, and so has developed a tactic to 'see' them as just other people. For instance, he recorded his parents' telephone numbers in his cell phone book with their names, rather than identifying them as mother and father. This way, 'when they call me, I read their name on the screen. So they are just like any other person who calls me'. He continues:

> When they call me, needless to say that they call me so often, I am as if I'm speaking with someone else. I mean normally, I get so stressed when my father is around, but when he calls me, of course, I am still stressed but not that much. For example, I smoke as I speak to my father on my cell, I tell lies easily that I cannot normally when we are face to face.

The absence-presence of cellular telephonic touch or coupling mediates relations between bodies in such a way that one may become freer and more comfortable when contacting another. As I mentioned in the previous chapter, some of my interviewees told me that they feel much more comfortable when they are texting to others, they feel the other as more contactable and touchable in some ways with cellular telephony (for similar accounts, see Pertierra, 2005; Goggin, 2006). Bodies may become more comfortable in telephonic interaction as well, more capable of manoeuvres that would be otherwise unthinkable in face-to face interaction (see also Taylor and Harper, 2003). The strange experience of co-presence through cellular telephony can liberate bodies momentarily in the course of coupling. As the other becomes a bodiless entity translated into text or voice, the intensity of love and trust may lessen alongside the intensity of fear, power, or control. Since cellular telephony's co-presence or touch is not identical to the corporeal, the content and flow of information and energy are different. Such coupling might induce new relations in which power and control, fear and insincerity are mitigated, just as it might induce a relation in which love and care are not felt as intensely as they are in corporeal co-presence.

In his critique of communication technologies, Albert Borgmann (2007) writes that even when people feel freer, such as seeming less shy when communicating via email, this does not mean that these people become more extraverted in their real life experiences as a result of their virtual experiences. Similarly, cellular telephonic coupling does not automatically enable people to feel more comfortable or less anxious in their corporeal experiences because of cellular telephonic coupling. Cellular telephony might not contain such power to immediately and obviously transform social reality – it might not create new and hitherto unimagined forms of social relations, just as it might not induce a decline in social capital and in democratic participation. Or perhaps the changes cellular telephony effects are happening more slowly than we can observe. In either case, the cell phone promises the sensation and affectivity of change; specifically in existing conditions of absence-presence, distance-proximity, or intimacy-distance.

Coupling with cellular telephony does not only describe a moment of spatiality where identities and social categories are affirmed, but also one in which the rigidities of the network are loosened and identities become fluid. The capacity to manoeuvre against modes of control can transform the cell phone into a machine that fools the authorities who recruit cellular telephony as an agent of surveillance. In this way, for better or worse, it may counteract the control and order of networks. For instance, the colonization of IMEI numbers (the 'unique' serial number of the machine) by replacing the original numbers with the same fake number makes it impossible to detect or track the owner of the cell phone. In Turkey, the PKK often uses cell phones as bombs, and connections or communications between PKK terrorists always occur via cell phone, because they have cloned their cell phones' IMEI numbers (Aksiyon, 2005).

Coupling thus retains the potential to exceed the network, to evade the means of control and identity. Moreover, as Sheller (2004) argues, this mobile communication system favours instant coupling and decoupling. Coupling and decoupling refers precisely 'to the capacity to manoeuvre across multiple social contexts that creates the gel-like character of contemporary communicational settings' (Sheller, 2004: 47). With a cell phone one feels connected, even when not coupled with anyone else (see Licoppe, 2004). Cellular telephony remains a contact for connection, for coupling as well as for decoupling. Each decoupling is an experience of decoupling from social context and from the other, and this encourages 'bodies to mix and blur into an inhomogeneous gel' (White, 1995: 12 cited in Sheller, 2004: 47). Just as coupling is a constant possibility, so too is decoupling. As Meyda,[28] who had recently moved to Ankara from the village of Sivas for her university education, said, she makes and receives calls or texts to different people continuously and simultaneously:

> Sometimes, I feel that I need some company. Most of the time, I just want to speak [to] somebody, it does not matter who this will be. Then I ring one of my friends. If that friend does not answer, I call another one. Sometimes I go through half of my telephone book and ring most of them. If I don't get any ring in return, though, I feel depressed.

If each coupling refers to a means of engaging with particular networks, then decoupling describes mobility across social networks. Decoupling indicates the possibility of connecting with others or de-linking from a given context and switching into another. When Meyda calls an old friend in Sivas, she experiences the travel to Sivas in the course of this coupling. Then she decouples with that friend and couples with another one, for instance, a friend from school in Ankara. As she couples and decouples, she slips from one identity to another, one social context to another, from one temporal focus to another. Each coupling is necessarily different from any other, just as each space is different from any other. In this way, cellular telephony becomes a medium of translation and improvisation.

A body is never just one 'person' that can be identified with a particular social category. One is many, as Deleuze and Guattari (2002) have told us. A contractor, Nihat,[29] in rural Ankara says:

> I don't know how many people I speak to on my cell during a day. I speak with workers, friends, my family and also prospective customers. I have many hats, I'm a father, a husband, a son, a boss, a seller, a neighbour, a friend…so there are a lot of people who need to contact with me [sic] and with whom I need to be in contact…as a result, I don't put down my cellular for a minute all day'.

By providing constant and easy means of coupling and decoupling, cellular telephony suggests that it is possible to sense and even put on multiple identities, through improvisational coupling and decoupling with others. Sheller argues that each body, what she terms 'person', is a node of 'story condensation' and 'identity' that 'occur[s] at the interface between multiple networks and strings of social organization' (2004: 45). Like Urry (2000; 2004), Sheller argues that mobile communication systems allow bodies to 'become more readily mobile through space because of the greater potential for "self-retrieval" at the other end of [the] journey' (ibid.). A body leaves traces of herself/himself in the cell phone's containing space – contact numbers, PIN numbers, account details, photographs, favourite songs – that 'allow them [to] more easily pick up various "story lines" through systems', and permits them to 'do things and talk to people without being present in a particular place, without even being in one place' (Sheller, 2004: 48). As a resource, cellular telephony thus exponentially increases the possibilities for easing in and out of contingent socialities and for picking up the multiple narratives through which a sense of individuality is experienced. Both coupling and decoupling facilitate the project of articulating individuality, a social face, and a singularity by offering easy and accessible detours through others.

Those who couple and decouple with each other become entities that are both absent/present, and these inhabitable positions of proximity and distance create a sort of crowd where each can enjoy solitude yet still feel connection. I suggest that to provide new or different forms of sociality, cellular telephony gathers bodies by promising the safe play of *fort/da* in crowds. The crowd here is not a physical collective body; rather it emerges out of the coupling and decoupling characteristic of cellular telephony. As such, it always has the potential to expand. Canetti writes that a crowd is either open and must grow to survive, or it is closed, which increases pressure on its permanence (1984: 17). Cellular telephony's crowd is an open crowd that must proliferate and spread in order to survive. The larger the crowd, the greater the pleasure in coupling and decoupling; the more open the crowd, the greater the possibilities for improvisations and translations. Thus an open crowd is integral to the play of *fort/da* with cellular telephony. In chapter one, following Tardean theory, I argued that cellular telephony has become social through imitation. The imitation of cellular telephony generates a sociality where coupling and decoupling themselves are imitated. Cellular telephony has become popular through imitation, and it has become a social practice through imitating its patterns of coupling and decoupling. I maintain that cellular telephony is integral to social struggles in social space. Social spaces wherein bodies are ranked based on their economic, social, or cultural capital, originate in the fear of the touch of the stranger (Canetti). I suggest that cellular telephony has become a specific social practice and collective attachment in Turkey in particular because it opens up a site where the imagination, sensation and experience of a crowd is possible. In this crowd bodies can play their own illusionary game of *fort/da*; the other's touch and uncanny co-presence are no longer fearful; each body feels connected to others and so part of the same larger force; these bodies sense and even see progress towards an illusory self-proper – in which the people of Turkey become a felt collective inhabiting the space of global cellular telephony where movement and mobility are ideally open to all. William Bogard writes of Canetti's crowd:

> He also knows that social order is a function of distance, i.e. differences in class, status, and power depend on smoothing machines that assign bodies a proper spacing, distribute their place in a matrix of possible locations, or possible zones of visibility and invisibility. (2000: 282)

Even in the crowd, in the collective unconscious movement that moves through the coupling and decoupling of bodies that ideally stand as a challenge to the social space in which melancholia prevails, there are social codes and hierarchies. The crowd of cellular telephony is more than social in its decoupling and in its fluidity, yet less than social in its network-like structure. Cellular telephony reterritorializes and deterritorializes the network. Importantly, the crowd of cellular telephony reterritorializes and deterritorializes the society.

BIBLIOGRAPHY

Adey, P. and Bevan, P. (2006). 'Between the Physical and Virtual: Connected Mobilities', M. Sheller and J. Urry (ed.) *Mobile Technologies of the City*, New York: Routledge.

Agar, J. (2004). *Constant Touch: A Brief History of the Mobile Phone*, Cambridge: Icon Books.

Agamben, G. (1993). *Stanzas: Word and Phantasm in Western Culture*, Minneapolis: University of Minnesota Press.

Agora, URL Document: [www.hurriyet.com.tr/agore/article_main.asp?si=5.] [accessed 01/05/2006].

Ahıska, M. (2003). 'Occidentalism: The Historical Fantasy of the Modern', Relocating the Fault Lines: Turkey beyond the East-West Divide, special volume, *South Atlantic Quarterly* 102: 2/3, Spring/Summer, pp. 351–379.

__________ (2005). *Radyonun Sihirli Kapısı: Garbiyatçılık ve Politik Öznellik*, Istanbul: Metis.

Ahmed, S. (2004a). 'Affective Economies' in *Social Text*, 22 (2), pp. 117–139.

__________ (2004b). 'Collective Feelings or the Impressions Left by Others', *Theory, Culture and Society*, 21 (2), pp. 25–42.

Akcan, E. (2005). 'Melankoli ve Öteki', Cogito 43 URL Document [http://www.eurozine.com/articles/article_2005-08-25-akcan-tr.html]

Akçam, T. (2004). *From Empire to Republic: Turkish Nationalism and Armenian Genocide*, London: Zed.

Aksiyon, vol. 541 [18 April 2005].

Aksoy, A. and Robins, K. (1994). 'Istanbul Between Civilization and Discontent', *New Perspectives on Turkey* 10, pp. 57–74.

__________ (1995). 'Istanbul Rising: Returning the Repressed to Urban Culture', *European Urban and Regional Studies* 2 (3), pp. 223–235.

__________ (1997). 'Peripheral visions: cultural industries and cultural identities in Turkey', *Environment and Planning*, 29, pp.1937–52.

Albertsen, N. and Diken, B. (2001). 'Mobility, Justification and the City', *Nordic Journal of Architectural Research*, 14 (1), pp. 13–21.

Alhassan, A. (2004). *Development communication policy and economic fundamentalism in Ghana*, Tampere, Finland: University of Tampere Press.

Althusser, L. (1971). 'Ideology and Ideological State Apparatuses', *Lenin and Philosophy and Other Essays*, London: NLB, pp. 121–177.

Altıntaş, B. (2003). *Mendile, Simite, Boyaya*, Istanbul: Iletişim.

Anderson, B. (2006). 'Becoming and being hopeful: towards a theory of affect', *Environment and Planning D: Society and Space*, vol. 24 (5), pp. 733–752.

Appadurai, A. (1990). 'Disjuncture and difference in the global cultural economy', M Featherstone (ed.), *Global culture: Nationalism, globalization and modernity*, London: Sage.

Beckmann, J. (2004). 'Mobility and Safety', *Theory, Culture and Society,* 21 (4-5), pp. 81–100.

Belge, M. (2007). *Bir Hayat… Tuba Çandar ile Söyleşi,* Istanbul: Do€an Press.

Berberoğlu, E. (1999). 'Cep Telefonu Kullanma Kursu', *Hürriyet*, [06 February 1999].

Best, M. L. (2003). 'The Wireless revolution and universal access' ITU (ed.), *Trends in Telecommunication Reform,* Geneva: ITU.

Bhabha, H. (1992). 'Postcolonial Authority and Postmodern Guilt', L. Grosberg, C. Nelson and P.A. Treichler (eds.), *Cultural Studies*, London: Routledge.

Bijker, W. E, Hughes, T. P, Pinch, T. (eds.) (1987). *The Social Construction of Technological Systems,* Cambridge: MIT Press.

Bijker, W. E and Law, J. (1992). *Shaping Technology/Building Society: Studies in Technological Change*, Cambridge: MIT Press.

Borch, C. (2005). 'Urban Imitations: Tarde's Sociology Revisited', *Theory, Culture and Society,* 22 (3), pp. 81–100.

Borgmann, A. (1984). *Technology and the Character of Contemporary Life: A Philosophical Inquiry*, Chicago: University of Chicago Press.

_________ (2006). 'Technology as a Cultural Force: For Alena and Griffin', *The Canadian Journal of Sociology*, 31, pp. 351–60.

_________ (2007). 'Information and reality at the turn of the century' in R. C. Scharff and V. Dusek, *Philosophy of Technology: The Technological Condition*, Oxford: Blackwell, pp. 571–578.

Bourdieu, P. (1990). *The Logic of Practice*, Oxford: Polity.

Bronfen, E. (1989). 'The Lady Vanishes: Sophie Freud and *Beyond the Pleasure Principle*', *South Atlantic Quarterly*, 88 (4), pp. 961–91.

Bulaç, A. (1998). 'Eğitim ve Teknoloji Kullanımında İrtica', *Zaman*, 1998 [1 April 1998].

Bull, M. (2004). 'Sound Connections: an aural epistemology of proximity and distance in urban culture', *Environment and Planning D: Society and Space*, 22.

Bursali, O. (2006). 'Devrim' *Bilim Teknik* (Cumhuriyet), pp. 3.

Butler, J. (1997). *The Psychic Life of Power: Theories of Subjection*, Stanford: Stanford University Press.

__________ (1999). 'Interview with Vikki Bell', *Theory, Culture and Society*, 16 (2), pp. 163–174.

__________ (2003). 'Afterword: After Loss, What Then' in D. Eng and D. Kazanjian, (eds.) *Loss: The Politics of Mourning*, California: University of California Press.

Callon M and Law, J. (2004). 'Guest Editorial: Introduction', *Environment and Planning D: Society and Space*, 22, pp. 3–11.

Canetti, E. (1984). *Crowds and Power*, New York: Farrar Strauss Giroux.

Castells, M. (2000). *The Rise of the Network Society, Vol. 1: The Information Age: Economy, Society and Culture*, Oxford: Blackwell.

Chakraborty, D. (2004). The Case of Mobile Phones in Sitakund, URL Document:[http://www.i4donline.net/issue/may04/sitakund_full.htm] [21. April. 2005].

Chakravarty, P. (2004). 'Telecom, national development and the Indian State: a postcolonial critique', *Media, Culture and Society*, 26 (2), pp. 227–249.

Chesher, C. (2007). 'Becoming the Milky Way: Mobile Phones and Actor Networks at a U2 Concert', *Continuum: Journal of Media and Cultural Studies*, 21 (2), pp. 217–41.

Cixious, H. and Clement, C. (1986). *The Newly Born Woman*, tr. by B. Wing, Minneapolis: University of Minneapolis Press.

Civaoğlu, G. (2001) 'Sesimi Duyan Var Mi?' *Milliyet* [19 August 2001]

Clark, N (2000). '"Botanizing on the asphalt": The Complex Life of Cosmopolitan Bodies', *Body and Society*, Vol 6 (3–4), pp. 12–33.

Comay, R (2006). 'The Sickness of Tradition', *Walter Benjamin and History*, Continuum Int.

Cooper, G., Green, N., Murtagh, G. and Harper, R. (2002). 'Mobile Society? Technology, Distance and Presence', S. Woolgar (ed.) *Virtual Society? Technology, Cyperbole, Reality*, Oxford University Press, Oxford, pp. 286–301.

Davison, R. (1990). 'The Advent of the Electric Telegraph in the Ottoman Empire', *Essays in Ottoman Turks and Turkish History, 1773 –1923*, Texas: University of Texas Press.

Deleuze, G. and Guattari, F. (2002). *A Thousand Plateaus: Capitalism and Schizophrenia*, Minneapolis: University of Minnesota Press.

Deleuze, G. (1994). *Difference and Repetition,* tr. by P. Patton, London: The Athlone Press.

Demir, T. (2005). *Türkiye'de Posta Telgraf ve Telefon Teşkilatının Tarihsel Gelişimi (1840–1920)*, Ankara: Posta Telgraf Teşkilatı Yayını.

Derrida, J. (1987). *The Postcard: From Socrates to Freud and Beyond*, Chicago: Chicago University Press.

Devrim, H. (2007). 'Cep Sayısını Biliyor musunuz?' *Radikal*, [18 January 2007]

Donner, J. (2005a). 'The Social and Economic Implications of Mobile Telephony in Rwanda: An Ownership/Access Typology', Glotz, P., Bertschi, S. and Locke, C. (eds.) *Thumb Culture: The Meaning of Mobile Phones in Society,* Transcript, Bielefeld, pp.89–100.

_________ (2005b). 'The rules of beeping: Exchanging messages using missed calls on mobile phones in sub-Saharan Africa', paper presented at Questioning the Dialogue: 55th Annual Conference of the International Communication Association, New York. (see also URL Document: [http://www.jonathandonner.com/donner-mobrev.pdf] [accessed on 21 August 2007].

Durgun, S. (2006). 'Homo Balkanicus, Cilgin Turkler ve digerleri', *Toplum ve Bilim*, 105, pp. 245–264.

Dündar, C. (2004). 'Derin Türkiye'nin Lideri' *Milliyet* [26 March 2004].

Edensor, T. (2002). 'Material Culture and National Identity', *National Identity in Popular Culture*, New York: Berg.

_________ (2004). 'Automobility and National Identity: Representation, Geography and Driving Practice', *Theory, Society and Culture*, 21 (4–5), pp. 101–120.

Ekonomik Trend, [05 February 2001].

Eickelman, D. (2004). 'Mobile Telephones as New Media in the Middle East and North Africa: Social and Economic Implications', paper presented at the Conference on Mobile Communication and Social Change, Korea, held on 18–19 October 2004.

Ellwood-Clayton, B. (2005). 'Desire and Loathing in the Cyber Philippines', in Harper, R., Palen, L. and Taylor, A. (eds.), *The Inside Text: Social, Cultural and Design Perspectives on SMS*, Springer, pp. 195–219.

Eng, D. and Kazanjian, D. (eds.) (2003). *Loss: The Politics of Mourning*, California: University of California Press.

Erdoğan, İ. (2001). 'Cep Telefonuyla Dolan ve Boşalan Cepler ve Bilinçler', *Bilim ve Ütopya*, 83, May.

Erener, S. (2001). 'Interview with Ahmet Tulgar' *Milliyet Pazar*, [23 December 2001].

Ergur, A. (2005). 'Kredi kartı kullanımında zaman algılaması ve borçlu yaşam' *Toplum ve Bilim*, 102, pp. 128–141.

Featherstone, M. (2004). 'Automobilities: An Introduction', *Theory, Society and Culture*, 21 (4–5), pp. 1–25.

Finansal Forum [10 December 2001].

Fischer, C. (1988). *America calling: A social history of the telephone to 1940*, Berkeley: University of California Press.

Fortunati, L. Katz, J.E. and Riccinin, R. (eds.) (2003). *Mediating the Human Body: Technology, Communications and Fashion*, New Jersey: Lawrence Earlbaum.

Fortunati, L. (2001). 'The mobile Phone: An Identity on the Move', *Personal and Ubiquitous Computing*, 5, pp. 85–98.

_________ (2005). 'Mobile Telephone and the Presentation of Self', R. Ling and P. Pederson (eds.) *Mobile Communications: Renegotiation of the Social Sphere*, London: Springer.

Freud, S. (1990). *Beyond the Pleasure Principle*, New York: Norton.

_________ (2001). 'Mourning and Melancholia', *Standard Edition of the Complete Pscyhological Works of Sigmund Freud, Vol. XIV (1914-1916)*, London: Vintage, pp. 243–259.

_________ (2002). *Civilization and Its Discontents,* London: Penguin.

Garcia-Montes, J., Caballers-Munoz, D. and Perez-Alvaroz, M. (2006). 'Changes in the self resulting from the use of mobile phones', *Media, Society and Culture*, 28 (1), pp. 67–82.

Geray, H. (2001). 'Toplumsal Değişkenler ve cep telefonu', *Bilim ve Ütopya*, 83, May.

_________ (2002). *İletişim ve Teknoloji: Uluslararasi Birikim Düzeninde Yeni Medya Politikalari,* Ankara: Ütopya Press.

Geser, H. (2005). 'Towards a Sociological Theory of the Mobile Phone', Zerdick, A., Picot, A., Scrape. K., Burgleman, J-C., Silverstone, R., Feldmann, V., Wernick, C. and Wolff, C. (eds.) *E-Merging Media: Communication and the Media Economy of the Future*, London: Springer, pp. 235–60.

Gilroy, P. (2004). *Postcolonial Melancholia*, New York: Columbia University Press.

Gittlin, T. (1987). *Watching television: A Pantheon Guide to Popular Culture*, Pantheon Press.

Glotz, P., Bertschi, S. and Locke, C. (eds.) (2005). *Thumb Culture: The Meaning of Mobile Phones in Society*, Bielefeld: Transcript.

Goankar, D. P. and Povinelli, E. A. (2003). 'Technologies of Public Forms: Circulations, Transfiguration, Recognition', *Public Culture*, 15 (3), pp. 385–397.

Goggin, G. (2006). *The Cell Phone Culture: Mobile Technology in Everyday Life*, London: Routledge.

Göle, N. (1991). *Medeniyet ve Örtünme*, Istanbul: Metis.

_________ (1992). *Modern Mahrem*, Istanbul: Metis.

_________ (1994). 'Towards an Autonomization of Politics and Civil Society in Turkey', Heper, M. (ed.) *Politics in the Third Turkish Republic*, Boulder: Westview Press.

_________ (1997). 'The Quest for the Islamic Self within the Context of Modernity', Kasaba, R. and Bozdoğan, S. (ed.) *Rethinking Modernity and National identity in Turkey*, Seattle: University of Washington Press, pp. 81–113.

_________ (2002). *Melez Desenler*, Istanbul: Metis.

Graham, S. and Marvin, S. (1996). *Telecommunications and the City: Electronic Spaces, Urban Places*, London: Routledge.

_________ (2001). *Splintering Urbanism: Networked Infrastructures, Technological Mobilities and the Urban Condition*, London: Routledge.

Green, N. (2002). 'On the Move: Technology, Mobility and the Mediation of Social Time and Space', *The Information Society*, 18 (4), pp. 281–292.

Grossberg, L. (1990). *We Gotta Get Out of This Place: Popular Conservatism and Postmodern Culture*, New York: Routledge.

Guattari, F. (1995). *Chaosophy*, New York: Semiotex(e).

Gürbilek, N. (2001). *Kötü Çocuk Türk*, Istanbul: Metis.

_________ (2003). 'Dandies and Originals: Authenticity, Belatedness, and the Turkish Novel', *The South Atlantic Quarterly*, Vol. 102 (2/3), pp. 599–628.

_________ (2007). *Vitrinde Yaşamak*, Istanbul: Metis.

Haddon, L. (2000). 'The Social Consequences of Mobile Telephony: Framing Questions', paper presented at the seminar of Sociale Konsevenser av mobiltelefoni, organized by Telebor, 16 June 2000, Oslo.

Hall, S. (1991). 'The Local and the Global: Globalization and Ethnicity' in A. D. King (ed.) *Culture, Globalization and the World-System*, Binghamton: State University of New York.

Haraway, D. (1992). 'The Promises of Monsters: A Regenerative Politics for Inappropriate/d Others', in Grossberg, L., Nelson, C. and Treichler, P. A. (eds.) *Cultural Studies*, New York: Routledge, pp. 295–337.

Harries, K. (1992). 'Context, Confrontation, Folly', *Perspecta*, 27, pp. 7–19.

Harkin, J. (2003). *Mobilization: The Growing Public Interest in Mobile Technology*, London: Demos.

Hansen, M. (2004). 'The Time of Affect, or Bearing Witness to Life' in *Critical Inquiry*, 30 (3), pp. 584–626.

Hayles, K. (2002). 'Flesh and Metal: Reconfiguring the Mindbody in Virtual Environments', *Configurations*, 10, pp. 297–320.

Heidegger, M. (1971a). 'Building, Dwelling, Thinking' in Hofstadter, A. (ed.) *Poetry, language, thought*, New York: Harper and Row.

_________ (1971b). 'The Thing' 'in Hofstadter, A. (ed.) *Poetry, language, thought*, New York: Harper and Row.

Hill, D. (2003). 'Communication for a new democracy: Indonesia's first on-line elections', *The Pacific Review*, 16 (4), pp. 525–548.

Hürses [12 November 2006].

Ihde, D. (1990). *Technology and the Lifeworld*, Bloomington: Indiana University Press.

Ito, M. (2005). 'Mobile Phones, Japanese Youth and the Replacement of Social Contact', Ling, R. and Pederson, P. (eds.) *Mobile Communications: Renegotiation of the Social Sphere*, London: Springer.

İnam, A. (2001). 'Düzenin Eli Cebimizde', *Bilim ve Ütopya*, 83, May.

Interpromedya and Birleşik Araştırmacılar (2007). 'The consumer behaviours of cellphone users in Turkey' URL Document: [www.kanal7.com.tr/ceptelefonu/arastirma] [accessed on 22 May 2007].

Johnston, A. (2004). 'The Cynic's Fetish: Slavoj Zizek and the dynamics of belief', *Journal for Psychoanalysis, Society and Culture*, 9 (3), pp. 259–71.

Jung, C. G. (1954). *The Psychology of the Transference*, Princeton: Princeton University Press.

Kadıoğlu, A. (1996). 'The Paradox of Turkish Nationalism and the Construction of Official Identity' in Kedourie, S. (ed.), *Turkey: Identity, Democracy, Politics*, London: Frank Cass.

Kafadar, C. (1995). *Between Two Worlds: The Construction of the Ottoman State*, Berkeley: University of California Press.

Kahraman, H. B. (2007). 'Interview with Neşe Düzel', *Radikal* [26 February 2007].

Kandiyoti, D. (1993). 'Identity and its Discontents: Women and the Nation', Williams, P. and Chrisman, L. (ed.), *Colonial Discourse and Post-Colonial Theory: A Reader*, New York: Harvester Wheatsheaf.

Kandiyoti, D. (1997). 'Gendering the modern: on missing dimensions in the study of Turkish modernity', Kasaba, R. and Bozdoğan, S. (eds.),

Rethinking Modernity and National Identity in Turkey, Seattle: University of Washington Press.

Kandiyoti, D. and Saktanber, A. (eds.) (2002). *Fragments of Culture: The Everyday of Modern Turkey*, London: I.B.Tauris.

Karaosmanoğlu, D. (2006). *Cooking The Past: The Revival of Ottoman Cuisine*, unpublished PhD Dissertation, McGill University.

Kasaba, R. and Bozdoğan, S. (eds.) (2000). *Rethinking Modernity and National Identity in Turkey*, Seattle: University of Washington Press.

Katz, J. (ed.) (2003). *Machines that Become Us: The Social Context of Personal Communication Technology*, New Jersey: Transaction Publishers.

Katz, J. and Aakhus, M. (eds.) (2002). *Perpetual Contact: Mobile Communication, Private Talk, Public Performance*, Cambridge: Cambridge University Press.

Katz, J. and Sugiyama, S. (2005). 'Mobile Phones as Fashion Statements: The Co-Creation of Mobile Communication's Public Meanings', Ling, R. and Pederson, P. (eds.) *Mobile Communications: Renegotiation of the Social Sphere,* London: Springer, pp. 63–84.

Kim, S. D. (2002). 'Korea: Personal Meanings', Katz, J. and Aakhus, R. (eds.) *Perpetual Contact: Mobile Communication, Private Talk, Public Performance,* Cambridge: Cambridge University Press, pp. 63–79.

Kittler, F. (1999). *Gramophone, Film, Typewriter*, Stanford: Stanford University Press.

Klein, N. (1997). *The History of Forgetting: Los Angeles and The Erasure of Memory*, London: Verso.

Kuzuluoğlu, S. (2007). *Radikal*, [10 September 2007].

Lallana, E. C. (2004). *SMS in business and government in the Phillipines,* Manila: DOST.

Larkin, B. (2004). 'Degraded Images, Distorted Sounds: Nigerian Video and the Infrastructure of Piracy', *Public Culture*, 16 (2), pp. 289–314.

Lasen, A. (2004). 'Affective technologies, emotions and mobile phone' URL Document: [http://receiver-vodafone.com/11] [accessed on 14 November 2005].

Lash, S. and Urry, J. (1994). *Economies of Signs and Space*, London: Sage.

Latour, B. (1996). *Aramis or the Love of Technology*, Massachusetts: University of Harvard Press.

_________ (2002). 'Gabriel Tarde and the End of the Social', Joyce, P. (eds.). *The Social in Question*, London: Routledge.

Law, J. and Mol, A. (1994). 'Regions, Networks and Fluids: Anaemia and Social Topology', *Social Studies of Science*, 24 (4), pp. 641–671.

_________ (2001). 'Situating technoscience: an inquiry into spatialities', *Environment and Planning D: Society and Space* 19 (5), pp. 609–621.

Lee, B. and LiPuma, E. (2002). 'Cultures of Circulation: The Imaginations of Modernity', *Public Culture*, 14 (1), pp. 191–213.

Lemish, D. and Cohen, A. (2005). 'Tell me how you use your Mobile and I'll tell you who you are: Israelis talk about themselves', Ling, R. and Pederson, P. (eds.) *Mobile Communications: Renegotiation of the Social Sphere,* London: Springer, pp. 187–202.

Levy, P. (1997). *Collective Intelligence: Toward an Anthropology of Cyberspace,* tr. Bononno, R., New York: Plenum.

Lewis, B. (1969). *The Emergence of Modern Turkey*, Oxford: Oxford University Press.

_________ (1997). *The Future of the Middle East*, London: Phoenix.

Licoppe, C. (2004). '"Connected" presence: the emergence of a new repertoire for managing social relationships in a changing communication technoscape', *Environment and Planning D: Society and Space,* 22, pp. 135–156.

Ling, R. and Yttri, B. (1999). '"Nobody sits at home and waits for the telephone to ring": Micro and hypercoordination through the use of the mobile phone', *Perpetual Contact Workshop*, Rutgers University, New Brunswick, N.J.

Ling, R. and Yttri, B. (2002). 'Hyper-Coordination via Mobile Phones in Norway', Katz, J. and Aakhus, M. (eds.) *Perpetual Contact: Mobile Communication, Private Talk, Public Performance*, Cambridge: Cambridge University Press.

Ling, R. (2003). 'Fashion and Vulgarity in the Adoption of the Mobile Phone among Teens in Norway', Fortunati, L. Katz, J. and Riccicin, R. (eds.) *Mediating the Human Body: Technology, Communication and Fashion,* Mahwah: Lawrence Earlbaum, pp. 93–102.

_________ (2004). *The Mobile Connection: The Cell Phone's Impact on Society*, Boston: Morgan Kauffmann.

Lobet-Maris, C. (2003). 'Mobile Phone Tribes: Youth and Social Identity', Fortunati, L., Katz, J. and Riccicin, R. (eds.) *Mediating the Human Body: Technology, Communication and Fashion*, Mahwah: Lawrence Earlbaum.

Lupton, D. (1998). *The Emotional Self: A Sociocultural Explanation,* London: Sage Publications.

Lloyd, D. (2003). 'The Memory of Hunger', Eng, D. and Kazanjian, D. (eds.) *Loss: The Politics of Mourning*, California: University of California Press.

Macherey, P. (2004). 'Out of Melancholia: Notes on Judith Butler's *The Psychic Life of Power: Theories in Subjection*', *Rethinking Marxism,* 16 (1), pp. 7–17

Mackenzie, A. (2004). 'From Café to Park Bench: Wi-Fi and Technological Overflows in the City', Sheller, M. and Urry, J. (eds.) *Mobile Technologies of the City*, New York: Routledge.

Mahçupyan, E. (2001). *Zaman* [19 July 2001].

_________ (2005). 'Millet Yoluyla Toplumun Askerileşmesi', *Zaman* [15 November 2005].

Mardin, Ş. (1990). *Türkiye'de Toplum ve Siyaset*, Istanbul: İletişim.

_________ (1994). 'Cultural Change and The Intellectual: A Study of The Effects of Secularization in Modern Turkey', Mardin, Ş. (ed.) *Cultural Transition in the Middle East*, Leiden: Social Economic and Political Studies of The Middle East, Vol. 48, pp. 153–189.

_________ (1997). 'Project as Methodology: Some Thoughts on Modern Turkish Social Science', in Bozsoğan, S. and Kasaba, R. (eds.) *Rethinking Modernity and National Identity in Turkey*, Seattle: University of Washington Press.

_________ (2005) 'Turkish Islamic Exceptionalism Yesterday and Today: Continuity, Rupture and Reconstruction in Operational Codes', *Turkish Studies,* 6 (2) pp. 145–165.

Massey, D. (1993). 'Power Geometry and a Progressive Sense of Place', Bird, T., Curtis, B., Putnam, T. and Robertson, G. (eds.) *Mapping the Futures: Local Culture, Global Change*, London: Routledge.

Massumi, B. (2002). *Parables for the Virtual: Movement, Affect, Sensation,* Durham: Duke University Press.

Mauss, M. (1979). *Sociology and Psychology*, London: Routledge.

Moon, M. (1995). 'Memorial Rags', in Haggerty, G. and Zimmerman, B. (eds.) *Professions of Desire: Lesbian and Gay Studies in Literature.* New York: Modern Language Association, pp. 233–40.

Miller, D. and Horst, H. (2005). '"Cell Phone comes like a blessing": Religion and the Cell Phone in a Rural Jamaican Town', *Jamaica Journal,* 29 (1–2), pp. 12–17.

Mumford, L. (1962). *Technics and Civilization.* New York: Harcourt Brace Jovanovich.

Mungan, M. (2000). 'Teknoloji ve Ideoloji', *Radikal İki*, [04 June 2000].

Munoz, J. E. (1999). *Disidentifications: Queers of Color and Performance of Politics,* Minneapolis: Minnesota University Press.

Nathanson, D. L. (1996). *Knowing Feeling: Affect, Script and Psychotherapy*, New York: Norton.

Navaro-Yashin, Y. (2002). *Faces of the State: Secularism and Public Life in Turkey*, Princeton: Princeton University Press.

Neyzi, L. (2001). '"Object or Subject" The Paradox of "Youth" in Turkey', *International Journal of Middle East Studies,* 33 (4), pp. 411–432.

Novak, P. (1999). 'Circles and Circles of Sorrow: In the Wake of Morrison's *Sula*', *PMLA*, 114, pp. 184–93

Nussbaum, M. (1996). 'Patriotism and Cosmopolitanism', *For the Love of Country: Debating the Limits of Patriotism*, Boston: Beacon Press.

Nyiri, K. (ed.) (2005). *A Sense of Place: The Global and the Local in Mobile Communication*, Vienna: Passagen Verlag.

Ogden, T. (1985). 'On Potential Space', *International Journal of Psycho-analysis*, 66, pp. 129–141.

_________ (1989). 'On the Concept of an Autistic-Contiguous Position' *International Journal of Psychoanalysis*, 70, pp. 127–140.

_________ (1992). 'The dialectically constituted/decentered subject of psychoanalysis II' *The Contributions of Klein and Winnicott,* International Journal of Psychoanalysis, 73, pp. 613–26.

_________ (2004). 'On holding and containing, being and dreaming', *International Journal of Psychoanalysis*, 85 (6), pp. 1349–1364.

Okabe, M., Ito D., Matsuda, M. (eds.) (2005). *Personal, Portable Pedestrian: Mobile Phones in Japanese Life*, Massachusetts: MIT Press.

Oksman, V. and Rautuaninen, P. (2003). 'Extension of the Hand: Children's and Teenagers' Relationship with the Mobile Phone in Finland', Fortunati, L., Katz, J. and Riccinin, R. (eds.) *Mediating the Human Body: Technology, Communications and Fashion*, Mahwah: Lawrence Earlbaum, pp.93–102.

Özbek M. (1997). 'Arabesk Culture: A Case of Modernization and Popular Identity', Kasaba, R. and Bozdoğan, S. (eds.) *Rethinking Modernity and National Identity in Turkey*, Seattle: University of Washington Press.

Özcan Y. Z. and Koçak, A. (2003). 'Research Note: A need or status symbol?' *European Journal of Communication*, 18 (2), pp. 241–254.

Özgüven, F. (1998). 'Türk Sinemasının Yükşeliş ve Çöküşü Üzerine Bazı Düşünceler', Dincer, S.M. (ed.) *Türk Sineması Üzerine Düşünceler*, Ankara: Doruk.

Pajnik, M. and Lesjak-Tusek, P. (2002). 'Observing discourses of advertising: Mobitel's interpellation of potential consumers', *Journal of Communication Inquiry*, 26 (3), pp. 277–299.

Pakkan, Ş. (2004). 'Cepten alo diyeli 10 sene oldu', *Milliyet Pazar*, [22 February 2004].

Pamuk, O. (2003). *Istanbul: Hatıralar ve Şehir*, Istanbul: Iletişim.

Paragas, F. (2005). 'Being Mobile with the Mobile Phone: Cellular Telephony and Renegotiations of Public Transport as Public Sphere' in Ling, R. and Pederson, P. (eds.) *Mobile Communications: Renegotiation of the Social Sphere*, London: Springer.

Park, W. K. (2005). 'Mobile phone addiction', Ling, R. and Pederson, P. (eds.) *Mobile Communications: Renegotiation of the Social Sphere*, London: Springer.

Parla, J. (2003). 'Car Narratives: A Subgenre in Turkish Novel Writing', *The South Atlantic Quarterly*, 102 (2–3), pp. 535–550.

Perry, M., O'Hara, K., Sellen, A., Brown, B. and Harper, R. (2001). 'Dealing with Mobility: Understanding Access Anytime, Anywhere', *Transactions on Computer Human Interaction*, 8 (4), pp. 323–347.

Pertierra, R. (2005). 'Mobile Phones, Identity and Discursive Intimacy', *Human Technology*, 1 (1), pp. 23–44.

Plant, S. (2000). 'On the Mobile: The Effect of Mobile Telephones on Social and Individual Life', URL document: [http://www.motorola.com/mot/documents/0,1028, 333, 00.pdf] [accessed on 1 April 2006].

Probyn, E. (2004). 'Everyday Shame', *Cultural Studies*, Vol. 18 (2/3), pp. 328–349.

Puro, J. K. (2002). 'Finland: a Mobile Culture', Katz, J. and Aakhus, M. (eds.) *Perpetual Contact: Mobile Communication, Private Talk, Public Performance*, Cambridge: Cambridge University Press, pp. 19–29.

Radden , J. (2000). *The Nature of Melancholy: From Aristotle to Kristeva*, Oxford: Oxford University Press.

Radikal Cumartesi [24 February 2007].

Rafael, V. L. (2003). 'The Cell Phone and the Crowd: Messianic Politics in the Contemporary Philippines', *Public Culture*, 15 (3), pp. 399–425.

Rajchman, J. (2000). *Constructions*, Boston, MA and London: MIT Press.

Rakow, L. and Navarro, V. (1993). 'Remote Mothering and the Parallel Shift: Women meet the cell phone', *Critical Studies in Mass Communication*, 10, pp. 144–157.

Ramazani, J. (1994). *Poetry of Mourning: The Modern Elegy from Hardy to Hearey*, Chicago: University of Chicago Press.

Reddy, W. M. (2001). *The Navigation of Feeling: A Framework for the History of Emotion*, Cambridge: Cambridge University Press.

Redfield, M. and Farrell, J. (eds.) (2002). *High Anxieties: Cultural Studies in Addiction*, California: University of California Press.

Rheingold, H. (2002). *Smart Mobs: The Next Social Revolution*, Cambridge: Perseus.

Richardson, I. (2007). 'Pocket Technoscape: The Bodily Incorporation of Mobile Media', *Continuum*, 21 (2), pp. 205–215.

Ronnel, A. (1989). *The Telephone Book: Technology, Schizophrenia, Electric Speech*, Nebraska: University of Nebraska Press.

_________ (1995). 'Our Narcotic Modernity', Conley, V. (ed.) *Rethinking Technologies*, Minnesota: University of Minnesota Press.

Rose, M. (2006). 'Gathering 'dreams of presence': a project for the cultural landscape', *Environment and Planning D: Society and Space,* 24, pp. 399–425.

Ross, K. (1995). *Fast Cars, Clean Bodies: Decolonization and Reordering of French Culture*, London: MIT Press.

Ross, S. D. (1998). *The Gift of Touch: Embodying The Good,* NY: SUNY Press.

Samuel, J., Shah, N. and Hadingham, W. (2005). 'Mobile communications in South Africa, Tanzania, and Egypt: Results from community and business surveys', URL Document: [http://www.vodafone.com/assets/files/en/AIMP_09032005.pdf] [accessed on 8 April 2006].

Santner, E. (1993). *Stranded Subjects: Mourning, Memory and Film in Post-War Germany,* Ithaca: Cornell University Press.

Sartre, J. P. (2001). *Basic Writings*, Priest, S. (ed.), London: Routledge.

Savage, S. (2003). 'Posting to smartmobs', URL Document: [http://www.smartmobs.com/archive/2003/08/10/rheinghold_media.html] [13 August 2003].

Schwartz, R. (1996). *Wireless communications in developing countries: Cellular and Satellite Systems,* Norwood, MA: Artech House Publishers.

Sconce, J. (2000). *Haunted Media: Electronic Presence from Telegraphy to Television,* Durham: Duke University Press.

Seltzer, M. (1992). *Bodies and Machines*, London: Routledge.

Serres, M. (1995). *Genesis*, Michigan: The University of Michigan Press.

Sheller, M. (2004a). 'Mobile publics: beyond the network perspective', *Environment and Planning D: Society and Space*, 22, pp. 39–52.

_________ (2004b). 'Automotive Emotions: Feeling the Car', *Theory, Society and Culture*, 21, pp. 4–5.

Sheller, M. and Urry, J. (2000). 'The city and the car', *International Journal of Urban and Regional Research,* 24, pp. 203–220.

_________ (2006a). 'Introduction: Mobile Cities, Urban Mobilities' in Sheller, M. and Urry, J. (eds.) *Mobile Technologies of the City,* New York: Routledge.

_________ (2006b). 'The New Mobilities Paradigm', *Environment and Planning A*, 38, pp. 207–226.

Skeggs, B. (2004). *Class, Self, Culture*, London: Routledge.

Sofia, Z. (1999). 'Machinic Musings with Mumford', *m/c journal*, 2 (6).

_________ (2000). 'Container Technologies', *Hypatia*, 15 (2), pp. 181–201.

Sorensen, C. (2005). 'Technology, time and everyday life', paper presented to Mobile Technologies Conference, Oxford: Oxford Internet Institute, 25 November 2005.

de Souza e Silva, A. (2006). 'Interfaces of Hybrid Spaces', Kavoori, A. and Arceneaux, N. (eds.) *The Cell Phone Reader: Essays in Social Transformation,* New York: Peter Land, pp. 19–30.

Sterne, J. (2003a). *The Audible Past: Cultural Origins of Sound Reproduction*, Durham: Duke University Press.

_________ (2003b). 'Bourdieu, Technique and Technology', *Cultural Studies,* Vol. 17 (3/4), pp. 367–389.

Stiegler, B. (2006). 'Disaffected Individuals' in URL Document: [http://www.arsindustrialis.org/Members/pcrogan/disaffectedindividual] [accessed on 1 June 2007].

Striphas, T. (2006). 'Translation', Shepherd, G.J., St. John, J. and Striphas, T. (eds.) *Communication as . . .: Perspectives on Theory*, London: Sage, pp. 232–241.

Strom, G. (2002). 'The telephone comes to a Filipino village', Katz, J.E. and Aakhus, M. (eds.) *Perpetual Contact: Mobile Communication, Private Talk, Public Performance*, Cambridge: Cambridge University Press, pp. 274–283.

Suner, A. (2001). 'Bir Bağlantı Koparma Aracı Olarak Türkiye'de Cep Telefonu', *Toplum ve Bilim*, Vol. 90.

Şinas, K. (2000). *Radikal Iki* [20 August 2009].

Tall, S. M. (2004). 'Senegalese émigrés: New information and communication technologies', *Review of African Political Economy*, 31 (99), pp. 31–48.

Tarde, G. (1962). *The Laws of Imitation,* Gloucester: Peter Smith.

Taylor, A. and Harper, R. (2003). '"The Gift of the Gab?" A Design Oriented Sociology of Young People Use's of Mobiles', *Computer Supported Co-operative Work,* 12 (3), pp. 267–296.

Taylor, A. (2005). 'Phone-Talk', Ling, R. and Pederson, P. (eds.) *Mobile Communications: Renegotiation of the Social Sphere*, London: Springer, pp. 149–166.

Tekeli, I. and Ilkin, S. (1992). 'Osmanli Imparatorluğunda 19. yyın Ikinci Yarısında Nafia Programları ve Teknoloji Üzerine', *Dünü ve Bugünüyle Toplum ve Ekonomi*, 3, April.

Tekelioğlu, O. (1996). 'The Rise of a Spontaneous Synthesis: The Historical Background of Turkish Popular Music', in Kedourie, S. (ed.) *Turkey: Identity, Democracy, Politics*, London: Frank Cass&Co, pp. 194–215.

Thrift, N. (1996). *Spatial Formations*, London: Sage.

__________ (2000). 'Still Life in Nearly Present Time: The Object of Nature', *Body and Society*, Vol 6 (3–4), pp. 34–57.

Townsend, A. (2001). 'Mobile Communication in the Twenty-First Century City', Brown, B., Green, N. and Harper, R. (eds.) *Wireless World: Social and Interactional Aspects of the Mobile Age*, London: Springer, pp. 62– 80.

Törenli, N. (2005). 'Türkiye'de Telefon Haberleşmesi Alaninda Yaşanan Dönüşüm: Neo-liberal Politikalar Eşliğinde Ulus Devletin Elinden Alinan Politika Araçları', *Kültür ve Iletişim*, 8 (1), pp. 11–40.

TÜBİTAK (1997). 'The Science and Technology Policy of Turkey', URL Document: [http://www.inovasyon.org/pdf/AYK.Sabanci.BiyotekEkim00.pdf] [accessed on 6 August 2000].

Türkiye Araştırmaları Literatür Dergisi, Türk Bilim Tarihi, Fall 2004 (4).

Türkiye Telekommunikasyon Kurumu (2006). 'Sabit ve Mobil Telefon Kullanıcıların Profili ve Eğilimlerinin Belirlenmesi', URL Document: [http://www.ttk.gov.tr] [accessed on 4 September 2007].

Uras, G. (2004). 'Simit parasını kredi kartıyla 3 taksitte ödeyeceğim', *Milliyet*. [16 July 2004].

Urry, J. (2000). *Sociology Beyond Societies*, London: Routledge.

__________ (2002). 'Mobility and Proximity', *Sociology*, 36 (2), pp. 254–274

__________ (2003). *Global Complexity*, Cambridge: Polity Press.

__________ (2004) 'Connections', *Environment and Planning D: Society and Space*, 22, pp. 27–37.

Verbeek, P. P. (2005). *What Things Do: Philosophical Reflections on Technology, Agency, and Design*, Penn State: Penn State Press.

Vershinskaya, O. (2002). *Mobile communication: Use of mobile phones as a social phenomenon: The Russian experience*. Revista de Estudious de Juventud, 52, pp. 139–149.

Watson, J. (1995). 'Guys and Dolls: Explatory Repetition and Maternal Subjectivity in the Fort/da Game', *American Imago*, 52 (4), pp. 463–503.

Weilenmann, A. and Larsson, C. (2001). 'Local Use and Sharing of Mobile Phones', Brown, B. Green, N. and Harper, R. (eds.) *Wireless World: Social and Interactional Aspects of the Mobile Age*, London: Springer, pp. 92–105.

White, H. (1992). *Identity and Control: A Structural Theory of Social Action*, Princeton: Princeton University Press.

Williams, S. J. (2001). *Emotions and Social Theory: Corporeal Reflections on the (Ir) rational*, London: Sage Publications.

Wills, D. (1995). *Prosthesis*, California: Stanford University Press.

Winner, L. (1993). 'Artifact/Ideas and Political Culture', Teich, A. (ed.) *Technology and the Future*, New York: St. Martin Press.

Winnicott D. W. (1965). 'The theory of the parent-infant relationship', *The maturational processes and the facilitating environment*, New York: International University Press.

_________ (1971). *Playing and Reality*, London: Routledge.

Yeni Şafak, [20 February 2001].

Yılmaz, M. A. (2004). 'Evlere Şenlik Bir Başbakan ile sohbet' yemeği', *Milliyet* [26 February 2001].

Yumul, A. and Özkırımlı, Ü. (2000). 'Reproducing the nation: "banal nationalism" in the Turkish press', *Media, Society, Culture*, 22, pp. 787–804.

Zhang, Y. and Harwood, J. (2004). 'Modernization and tradition in an age of globalization: Cultural values in Chinese television commercials', *Journal of Communication*, 54, pp. 156-172.

Zizek, S (2000) 'Melancholy and the Act', Critical Inquiry, 26 (4), pp. 657–681.

NOTES

1. INTRODUCTION: CELLULAR TELEPHONY, IMITATION, ATTACHMENT

1 See the 2004 Alcatel report on Turkey in the Alcatel Telecommunications Review. The report states that Turkey's mobile telecommunication market has the capacity to lead Europe in the 'penetration of mobile services as the young population takes on board new technologies and services' (320). See also the weekly journal *Ekonomik Trend*: 'As Turkish society has been so eager to adopt mobile technology of the global world, the global mobile industry has turned their attention to the Turkish market. 2001 will be a year for grand investments by the global mobile companies in Turkey' (05/02/2001). Also see Yeni Şafak's news story which notes, 'According to the research sponsored by German Deutche Bank, Turkish GSM sector is poised to become the most expansive market in the next five years despite the major economic crises' [20 February 2001]. The CEO of Turkcell, Turkey's market leader GSM provider in Turkey, comments: 'The growth of Turkcell's share in the market in terms of gaining more and more users especially in the third quarter of this year, in spite of the country's economic depression and the increasing competitiveness in the market with the introduction of new GSM provider companies, shows that mobile communication has been extremely appealing to Turkish society and it also shows that the market has the potential to expand more and more' (Oksan Sanor, *Finansal Forum*, 10 November 2001, p. 7) (my translation).

2 Industry heavyweights like Koc and Sabanci Holdings attempted to enter the cell phone market in 2001.

3 According to the 2006 research by TTK (Turkish Telecommunication Association), which surveyed 4322 houses across 81 cities, there are on average two cell phone users in each house. Among all interviewees, 82 per cent declared that they have at least one cell phone in their house

that is shared by their family. This research also reveals that mobile telephony is more popular than the wired telephone system. (URL document: http://www.ttk.gov.tr)

4 See http://www.tk.gov.tr/Yayin/istatistikler/istatistik/2010/gsm2010.htm

5 See M. Y. Yilmaz's column in *Milliyet*, 2004 [published on 26 February].

6 Avital Ronell (1989, 1995) whose work on the telephone and on technology blends a Heideggerian understanding of technology with a Derridean reading of postal movement, -along with Freud's theory of *fort/da* has been profoundly influential on my own search for understanding cellular telephony in Turkey. Thinking of cellular telephony as a response to a call of the collective is drawn specifically from Ronell's reading of technology.

7 Gabriel Tarde, a controversial figure in sociology, developed a cosmological theory of universal repetition, from which he extracts the matter that constitutes the social and imitation (1962). His theory has been reviewed by Bruno Latour (2002), who regards Tarde's work as a precursor of actor-network theory, and by Gilles Deleuze (1994), for whom Tarde's concept of imitation shows the unity of repetition and difference.

8 Bella Elwood-Clayton (2005) also notes that the issue of affordability is ironical in the sense that relatively low-income users choose to pay significant proportions of their income on cell phones and on texting, compared to other household goods or even vital expenses (200).

2. TECHNOLOGY OF/IN MAKING A MODERN NATION:MELANCHOLIC CONSTRUCT, MELANCHOLIC BODIES

1 Pamuk often describes the landscape of Turkey as one where the feeling and the state of being 'poor' determine ways of being, moving, and acting.

2 According to Jennifer Radden (2000), it was only after the nineteenth century that the literature on melancholia approaches this psychological phenomenon by differentiating the pathological disorder of melancholia from the emotional state or the mood of melancholy. In his book on Istanbul, Pamuk (2003) uses the Arabic word, *hüzün*, which is also used in Turkish to describe the melancholy of the landscape. He insightfully applies this term to the collective feeling of Istanbul's inhabitants in the aftermath of the foundation of the new republic. Although the difference between melancholia and melancholy is always

blurred in definitions, throughout this work I will use the term melancholia to refer to the psychic and social conditions of the people of Turkey, while using melancholy to reference the collective feeling of the people of Turkey.

3 Ahmet Hamdı Tanpınar immediately springs to mind when one thinks about melancholia in the historical context of Turkey. A novelist writing during Turkey's transition from the Ottoman Empire to the Turkish Republic, his works are often cast as manifestations of a deep melancholia for the landscape as a space of experience and imagination. Pamuk (2003), for instance, writes that Tanpınar was a modernist in a sense that he admired the reforms of M. Kemal Atatürk (the founder of the Turkish Republic), and was loyal to the new republic's project of modernization. Yet Tanpınar was also deeply melancholic because of the painful consciousness unleashed by losing the past.

4 Akcam specifically refers to the 1915 massacre of Armenian people in the Ottoman Empire. In the early 1900s, the Ottoman Empire was rapidly disintegrating – losing wars to the Europeans, and so losing its imperial power and territory. The new Turkish Republic was founded in 1923 by a group of Republicanists from the army, under the leadership of M. Kemal Atatürk. M. Kemal won the War of Independence against both the invading European countries and the Ottoman dynasty, which had collaborated with the European imperialist powers.

5 Here, it is significant to note that my understanding of melancholia opposes popular arguments that consider melancholia as a resistant aspect of identity politics, one which brings the possibility of insurrectionary activities to the forefront of everyday life (see Munoz, 1991; Ramazani, 1994; Bhabba, 1992; Moon, 1995; Novak, 1999). On the contrary I consider melancholia as a project of the body, which participates in the violent work of power, and which in one way or another generates losses on the part of its subject. In that respect, I agree with Butler who tells us that melancholia becomes a means of resistance as it turns into mourning, which as a potentiality always resides in the melancholic condition of the body (see also Santner (1993) for the discussion of successful mourning).

6 In Althusser's (1971) essay 'Ideology and Ideological Apparatuses', the constitution of the subject takes place through language, as the effect of the authoritative voice that hails the body. In the social scene that he depicts, an officer of the 'Law' hails from behind: 'hey, you there', an individual turns round, 'believing/suspecting knowing that it is for him, i.e. recognizing that "it really is he" who is meant by the hailing'

(118). The ideology succeeds in transforming a body to a subject when it makes the body to turn round and accept the interpellation of ideology. In other words, with the acceptance of the call of the law or an entrance into the language of self-ascription – 'here, I am' – the body becomes a subject.

7 I want to underline the significance of the work of power and the notion of contingency in generating melancholic subjects. Current literature defines melancholia in a number of different ways, from different perspectives and for different purposes. Some studies argue that in Freud's writings melancholia seems to describe the conditions through which the body becomes a subject; Freud writes in his later essays (such as 'On Narcissism', 'The Ego and Id') that the loss divides the ego into two parts, the ego and critical agency mentioned above (see Agamben, 1993, Comay, 2006). Butler's theory is in line with this argument. Yet other work on melancholia and subjectivity argues that melancholia is an interpretation of the subject's ontological lack, which is free from social and political foundations - an outcome of one's presence in the world as a human being that can exist insofar as it enters into the domain of language (Zizek, 2000). Butler's theory differs from these studies at this point, as her emphasis on the social production of subjectivity necessarily rejects the assumption that what makes the subject is separate from its contingent social, historical and political conditions. She writes, on the contrary, that the subject is the very relation of the body with its contingent conditions; in that sense its melancholia is always produced through the work of power whose given categories wound bodies. By stressing the cultural, social and political formation of melancholic subjects, Butler suggests that 'mourning' losses is possible. Rather than understanding melancholia as an outcome of ontological lack, she recommends we consider losses as historical and contingent, not all of which are immutable yet still require the work of mourning (see also Santner, 1996). Both Santner and Butler show us that what makes certain losses unrecognizable is power itself, not the nature of melancholia. When the social and political conditions needed for mourning are missing, melancholia sets in as a cultural form, precluding sociality and occluding the ability to imagine otherwise.

8 Anadolu refers to the landscape of rural Turkey, as a territory, people and 'culture'.

9 The term 'habitus' was first suggested by Mauss to describe the social nature of the body. Pierre Bourdieu's social theory is largely based on

understanding and theorizing the "habitus" of individual agents in relation to their society. The next chapter will discuss Bourdieu's theory in relation to technological practices. However, here we should note how Mauss understood this term. He wrote in his famous 'Techniques of the Body': 'The word translates better than "habitude" (habit and custom), the "exis" of Aristotle (who was a psychologist). It does not designate those metaphysical habitudes, that mysterious "memory", the subjects of volumes or short and famous theses. These "habits" do not just vary with individuals and their imitations, they vary especially between societies, educations, prestige. In them we should see the techniques and work of the collective and individual practical reason rather than, in the ordinarily [sic] way, merely the soul and its repetitive faculties' (1973: 73).

10 Ziya Gökalp had said, 'nationalism is not based on genealogy. It is based on a national ideal and on national education' (cited in Neyzi, 2001: 416).

11 Ahıska describes the state power in the 1930s through the 1940s as a hegemonic power exercised through state bureaucrats, army and elites. She also draws attention to the work of power in bodies or how 'objects' participate in forming a national fantasy – Occidentalism – as they become 'subjects' of the project. In this respect, the work of power can be thought of as hegemonic power to the extent that it generates consent among different agents. Yet if we speak of the bifurcation of reality into two parts based on the unspoken but known rules about what can and cannot be said, we need to emphasize the military, authoritarian aspect of the state power.

12 For Butler, homosexual love is lost as an outcome of subordination to social categories like gender. To gain the possibility of social recognition, a body loses its potential for homosexual love and as an outcome of the subjectification process, melancholy gender is instituted.

13 Paul Gilroy (2004) also provides a similar analysis of melancholia in British state/society on the basis of the loss of former imperial power.

14 We have seen above that the ideals of the Turkish Republic, as defined and imagined by the early Republicanists, echoed former idealizations by the Ottoman intellectuals. One can claim a longer continuity by linking some discursive elements that define ideals to even earlier periods (see Mardin, 1990). Thus we can fairly argue that the Turkish Republic's national ideals were based on a fantasy of actualizing the ghostly remains of the lost empire.

15 Here, one can see the influence of Avital Ronell's understanding of technology (in her reading of Heidegger and Freud) in my treatment of the social construction of technologies in the social imaginary in Turkey. However, rather than interpreting the 'guilt' as a metaphysical concept, I am trying to consider it in the historical, social, cultural and affective context.

16 Social scientists offer many observations about urban crimes from the 1990s onward. One of these crimes is what we call *kap-kac* ('grab-run') meaning that in public spaces or streets, a robber grabs handbags or purses and runs away, sometimes dragging the victim down the street. Some authors point out that criminals of this sort are generally street children – most are children from large Kurdish families that have been exiled from their villages due to the forced migration policies in the southeast region where a mostly Kurdish population lives and where the battle between the Turkish army and the terrorist group PKK has been taking place. In her book on street children of Ankara, Betül Altıntaş (2003) tells us that most street children who are pushed to crime (robbery mostly), either by their families or some other groups, have been exiled from their villages in the southeast part of Turkey. See also Gürbilek (2001), who speaks of the transformation of the image of the child in the social imaginary, from being an innocent to a criminal figure. The streets of big cities have become increasingly unsafe to wander around due to the increasing crime rates of *kap-kac* and other sort of robbery in the 1990s (including robbery of cell phones). Urban spaces, especially crowded spots, are now also potential places for terrorist bombings by the PKK and the Turkish groups linked to Al-Qaeda and Hizbullah. These terrorist groups often use the cell phone as a container for bombs.

17 The Turkish state has been working towards membership in the European Union. In 2003 official negotiations between Turkey and the EU began. The EU, however, required that Turkey make a set of political decisions that contradicted the country's internal and international policies. This tense relation with the European Union has generated a significant sense of frustration on the part of Turkish people, and provoked a feeling of nationalism in its contemporary political sphere.

18 The 2001 film *Şellale*, directed by Semir Aslanyürek, powerfully depicts this polarization, dramatizing the rift between left-wing party supporters and right-wing party supporters living in a village in the south part of the country.

19 The recent film *Babam ve Oğlum*, directed by Çağan Irmak (2005) emphasizes the violence of military state power against social bodies during the coup period through a generational story of a grandfather, father and his son.

20 I rely on a documentary about the last coup, *12 Eylül*, produced by Mehmet Ali Birand, Can Dündar and their crew for Channel D in 2000.

21 In the aforementioned documentary, *12 Eylül*, it has also been said that many had supported this coup. Also *Hürriyet*, columnist Ertuğrul Özkök writes that many had been willing for the military to take over the government and end the street violence of the time.

22 Fatih Özgüven writes about this very in-betweenness which has finally become decipherable: 'On the other hand in these weird times in this weird space, our weird dwelling in this "in-betweenness"...We neither belong to here nor to there. And the probability of us belonging to everywhere... Yet our denial of this very in-betweenness... although the schizophrenia of this weirdness is felt at every moment in our life, we are unwilling to recognize it, we are unwilling to be fed by it. Although it could be the only key – for centuries perhaps – which would make us "us", we still do not comprehend it' (Fatih Özgüven, 1996: 348) (my translation).

23 Recent attempts by academics, journalists, and novelists to start a public discussion about the Armenian genocide or expulsion have all been crushed by various types of violence. For instance, Hrant Dink, an Armenian Turkish citizen, who wrote an essay about the Turkish governments' historical hostility against the Armenian Turks has been sentenced to jail. While he was planning to take the decision of the Turkish Court to the European Human Rights Court, he was murdered by a young ultra-nationalist who was apparently recruited by the illegal part of the state organ, what we call the 'deep state' in Turkey. After Dink's assassination, thousands of people took the streets in a show of collective outrage against state-run racism in Turkey. Yet, nothing has changed in terms of laws or state policies, and speaking publicly about the Armenian genocide is still an act that may bring severe legal sanctions.

24 That is to say, objects function for melancholic bodies in that they do the work of mourning on behalf of bodies. Things that come to lodge in the place of the loss commemorate and mourn for the loss. Zizek's discussion of externalizing 'duty' to machines, things or other bodies is a good example; canned laughter, for instance, relieves us of our duty

to laugh as it is laughing instead of us (1989: 35). In a similar manner, Agamben (1993) seems to argue that things or objects mourn for the loss, instead of the melancholic himself.

25 Hasan Bülent Kahraman (2007), a columnist and a scholar, recently commented to Neşe Düzel that 'society in Turkey has fallen ill. It has no relation with emotions, knowledge and reality, any more. It is a society, which lives in phantasmagoria, which does not show any curiosity for anything other than (locally produced) television dramas, which has stuck in the mud of pornography...from the family to state bureaucracy, this society produces and experiences violence' (*Radikal*, 2007) [published on 26 February].

26 Güngör Uras, a prominent economist and columnist for the daily newspaper *Milliyet*, suggests that the people of Turkey want to live 'life' yet lack the economic capability to do so. Credits cards and bank credits provide them with what they lack, yet people seem to have forgotten that they need to pay it back. He writes that consumption behaviour with regards to credit cards and bank credits in Turkey is based on a typical sentiment of '*yarın ola, hayrola*' which can be translated as 'let's see what goods tomorrow will bring'. (*Milliyet*, 2004) [published on 16 July].

3. RETHINKING THE TECHNOSCAPE AND CONTEXTUALIZING CELLULAR TELEPHONY IN TURKEY

1 Sara Ahmed's conception of globality emphasizes the asymmetrical relations of different groups or collectives to 'movement' (see Ahmed, 2004a; 2004b).

2 Here I rely on Grossberg (1990), who wrote, 'affect is the plane or mechanism of belonging and identification (of which identity, constructed through either ideological or psychoanalytical interpellations, is only one form)...affect defines a structure and economy of belonging' (84).

3 Affectation is often discussed in philosophy within the debates around subjectivity. For instance, Deleuze and Guattari (2002) say that forms of subjectivity emerge by being affected with the material world.

4 Both Süleymaniye and Çamlıca are different districts of Istanbul, where Bihruz's story takes place.

5 Bihruz has an accident in the end – his beloved car is crashed along with his dreams of becoming someone else.

6 See Vasıf Kortun's (an art critic) interview with Ahmet Öğün who presented an artwork on Devrim, titled 'Devrim, the failed car', in his Zagrep exhibition. URL Document: [http://www.resmigorus.blog-spot/2007/03/interview] [accessed on 1 June 2007].

7 Nurdan Gürbilek (2003) uses this phrase to describe the mentality of Turkish novelists who approached society as 'lacking' the essential means to reconcile modernity with what is our own.

8 See for instance, Erol Manisalı's column in *Cumhuriyet*, 2001 [published on 13. August]; Ege Cansen's column in *Hürriyet*, 1998, [published on 25. April]; Günlük Evrensel, 2003 [published on 14 May] p. 2; Birgün, 2005 [published on 26 December] p.7.

9 I would like to highlight that Nurdan Gürbilek speaks about the 'belatedness' in relation to the literary genre of the novel as a product of the west, which was imitated in Turkey relatively later. However her analysis is perfectly applicable to a historical analysis of the techno-scape of Turkey for the reasons that I depict above.

10 Turgut Özal's statement was highlighted in a newspaper advertisement for an exhibition entitled 'Lifestyles of Turkey from the 80s to 2000s' which was organized by Meltem Ahıska, Zafer Yenal and Bülent Erkmen with the sponsorship of Osmanlı Bankası in 2005.

11 Irene Markoff writes, 'In an effort to combat the staggering commercial success of Arabesque, the state has been struggling to retaliate with subtle policies that aim to dethrone its stars by shifting attention to Western popular modes of expression. In promoting Western popular music on state-controlled airwaves, the state hopes to create a market that will compete with the Arabesque audio- and video-cassette industry. It is hoped that these and other measures will help lessen the potential of Arabesque to promote a Turkish national identity that identifies more with the backwardness of the Islamic "East" than with the progressive and modern "West" (1994: 228).

12 People who used wireless technology call themselves 'friends of wireless'.

13 '*Arkadaş arıyorum, arkadaş*' in Turkish.

14 Before then ultra-nationalism was voiced by some newspapers (for instance *Ortadoğu*), ruled by the ultra-nationalist party (MHP). Now, however, the voice of ultra-nationalism is represented by popular newspapers and television channels. We also need to stress as we briefly speak about the mediascape that broadcasting in any other language, such as Kurdish or Armenian, was prohibited until the process of negotiation with and constitutional adoption of the European Union had started.

4. ATTACHMENT TO CELLULAR TELEPHONY: THINKING OF MEANING, FUNCTION AND BODILY RELATIONS

1 *Radikal*, 2007 [published on 24 February] p. 3. Taken from a published interview with a street perfume seller in Istanbul in *Radikal.*

2 Meryem is a university student, who moved to Ankara very recently from a village im Amasya for her education. She comes from a peasant family and receives a municipal and governmental financial aid. I conducted an interview with her in Ankara, on 23 June 2007.

3 Here I rely on Marc Redfield and Janet Farrell's review of how addiction is conceptualized in theoretical works and in popular discourses (see their *High Anxieties: Cultural Studies in Addiction*, 2002). They maintain that the word 'addiction' obtained its meaning of 'compulsive habit' in the twentieth century, when the use of opium and similar drugs was defined within a new scientific paradigm. The word addiction, then, connotes a negative judgment of the addict's behaviour, a condemnation of supposed excess. My intention here is not to pass moralistic judgment about collective addiction to cellular telephony in Turkey. Rather, I am attempting to understand how and why cellular telephony becomes an object of collective attachment and addiction, and in particular to grasp the possible pleasures of the body it promises.

4 The ways in which I understand the concept of articulation will become clearer in chapter five, however, I shall briefly note that when technology is understood as a form of articulation, it follows that it is a gathering, a jointing of many different parts such as machines, practices, bodies that use it, performances of technology, conditions of possibility, etc. which is to say it does not only consist of its instrumental and symbolic value.

5 After writing that the 'technological artifact does not have an essence, an identity in itself, it becomes what it is in relation to people', Don Ihde notes that once technologies have received a (relative) identity '...within that relation they nevertheless can have their own weight' (1990) which is to say technologies co-shape the use that is being made of them.

6 See chapter two and three, where I discuss Nurdan Gürbilek's description of the 'third worlds' of Turkey in reference to the 'others' of ideal national identity such as Kurds, newcomers to the city, the peripheral, Islamists and also the 'other' of the urbanite self which has been repressed under the strict politics and modernization and secularization policies of the state.

7 See for scholarly works, Özcan and Koçak, 2003; Geray, 2001; and for journalistic representations Çetin Altan, 2004 in *Milliyet* [published on 20 November], Ece Temelkuran, 2001 in *Milliyet* [published on 17 December], Türker Alkan, 2007 in *Radikal* [published on 19 January], Durmuş Hocaoğlu, 2001 in *Aksiyon* [published on 12 January]; Doğan Kuban in *Cumhuriyet*, 2001; M. Y. Yılmaz in *Milliyet*, 2004 [published on 27 February]; M. Ş. Evgi in *Milli Gazete*, 2002 [published on 2 March].

8 *Gecekondu* literally means a construct that is built at night. One floor, small and illegal constructs are named as *gecekondu* in Turkey where big cities have been experiencing big flows of internal migration from rural areas. It is said that now in big cities, 60 per cent of urban spaces is filled with gecekondu districts. These places are traditionally labelled as the periphery of Turkey in strict association with poverty, ignorance, violence, etc. in the elitist discourses.

9 Cemil is Kamil's brother-in-law. He was generally silent in our talk, only occasionally nodding in agreement with Kamil, or by saying a few words to add detail to Kamil's statements. He used to work as an illegal street vendor, selling and repairing watches. He told me that people are no longer willing to get their watches repaired, they prefer to buy a new one. To feed his family, he needed some other work, and the boom in the cell phone market has 'saved his life and his family'. I conducted this interview with Kamil and Cemil on 25 June 2007 in Ankara.

10 In Turkey, every male Turkish citizen over 18 years old has to serve in the military for a time determined by his educational background. For instance a man who does not hold a university diploma has to serve the longest period, while others' (e.g. those who have university or masters degrees) service is shorter. Kamil served a year and a half because he dropped out of high school when he was 15. Thus educational qualifications (i.e. cultural capital) is important for a male Turkish citizen and play a formative role in ranking his position in the military.

11 Just like Kamil, most of my informants told me that they bought their first cell phones because they thought it was 'cool' to have a cellular, because they just 'wanted' to have it, because everybody around them had one, et cetera.

12 According to the newspaper, *Hürses*, Turkey's second-hand market is the largest in the world [published on 16 November 2006] (p. 9). I learned from Kamil that some cell phone shops only sell stolen cell phones. Kamil told me that it is so easy to sell a stolen cellular, because 'the state could not regulate it…they said that every owner has to fill out a form and have their machine recorded in files…but this did not

prevent robbery...now thieves just go to the graveyard and pick one and write his name, surname, birth date and the dead becomes the new owner...so the cell becomes recorded...and the police can't find this owner, because he is just dead...it is that easy...they need to find another way to stop this, but they just don't care...this is that type of country...you know' (25 June 2007).

13 Many columns, essays, and articles in newspapers and in journals argue that the popularity of cellular telephony in Turkey shows how easily the people of Turkey can adapt to modern technologies (Meral Tamer, *Milliyet*, 2001 [published on 4 May], Tamer, *Milliyet*, 2004 [published on 26 October], Abbas Güçlü, *Radikal*, 2004 [published on 16. November]). Although this may sound like expression of national pride, in most cases the popularity of cellular telephony is considered as part of our 'inferiority complex' or 'national lack' (for the reference matter see footnote number 55).

14 This is a common habit of cell phone users in Turkey. Fortunati (2000) observes the same bodily habit in Italy.

15 See Donner, 2005a for similar observations in relation to the cell phone as a product which provides the means of agency to its users in Rwanda.

16 Bulaç prefers to define the collective not as Islamists but as conservative Muslims.

17 Emine is 30 years old, female and living in Ankara. She is the owner of a local advertising agency which has a small work-load. She also works as a part-time manager at a restaurant/bar at nights. I have conducted two interviews with her. The first one was on 22 June 2007 in Ankara and the second one took place in the restaurant where she works, with her boss, Süreyya and Süreyya's husband, Mehmet on 1 August 2007.

18 Eda is a high-school teenager. She comes from a lower-class family and lives in one of the poorer districts of Ankara. She has participated in a focus group conversation whivh three other high-school teenagers attended on 24 June 2007, in Ankara.

19 Kamil tells me that this is one of most common reason for people.

20 See Elwood-Clayton, 2005 for similar observations in the Philippines.

21 She looks like a western traveller with her cowboy hat, mini-shorts and boots and wanders around the east part of Turkey where local women are in general veiled and are not very mobile. In the series of ads where the 'free girl' (*özgür kız*) walks through the small villages, we do not see any scene where she interacts with a local woman. In a number of scenes, she runs into local people, a young boy, an old peasant but never a local woman.

22 See *Sabah Pazar*, 2001 [16 December], p. 15, *Türkiye* 2001 [18 December], p. 12, *Cumhuriyet Dergi*, 2001 [30 December], p. 7, *Radikal Iki*, 2001 [04 March; 16 October, 19 December], p. 13; p. 7; p. 6, *Radikal Iki*, 2004 [25 January], p. 6, *Yeni Evrensel*, 2001 [02 January], p. 3.

23 *Milliyet Pazar*, 2001, an interview with Ahmet Tulgar.

24 Although people are now aware that it may be a threat to human health and is also polluting the earth.

25 According to TTK's 2006 research, cell phone users in Turkey change their cell phones in every two years on average. The same research states that 43 per cent of users have never changed their cell phones since they first purchased one. The majority of users (53 per cent) who have changed their cell phones said that they have changed it because the machine was broken or became obsolete. See URL Document: [www.ttk.org.tr].

26 Kaya is a 32 year old married man. He works in the public relations department at one of the Holdings of Turkey. I conducted two interviews with him. This quote was taken from the second interview which took place in Ankara, on 3 July 2007.

27 See for instance research conducted by Interpromedya and Birlesik Arastirmacilar in 2007. Their field analysis of technology users' habits with regards to cell phones, with a sample of 1338 users living in 16 big cities in Turkey, showed that people in Turkey use this technology for: telephoning (37 per cent), messaging (27 per cent), taking pictures (11 per cent), viewing videos and listening to music (7 per cent), audio recording (5 per cent), MMS (4 per cent), and emailing (2 per cent).

28 http://sozluk.sourtimes.org (see the 62nd entry by cybill in 'the cell phone' section).

29 Ronell writes in her *Narcotic Modernity*, 'science will still let you die' (1995).

30 I rely largely on my own viewing experiences of news programs on Show TV and Channel D in 2007. On one particular occasion, Show TV's news program chronicled a traffic accident that claimed the life of a middle-aged man. As the camera captured the scene, the dead body's cell phone rang. After hesitating for some time, a police officer answered the phone, and the caller happened to be the victim's daughter. The officer couldn't say anything about her father's death, only that he had been in an accident.

31 Paragas, who has done extensive studies on cellular telephony, specifically in the Philippines, offers a similar observation. He suggests that in the Philippines the cell phone has become something to protect

from thieves and conceal from others, for instance, by putting in silent mode, or by hiding it in a pocket so as not to attract attention (2005: 122).

32 Ahmet is a 47 years old taxi-driver, living in Ankara. I conducted an interview with him on 27 June 2007.

33 Emre is a high-school student at one of the public schools of Ankara. He participated in one of the focus group debates with other high-school teenagers on 28 June 2007 in Ankara.

34 I spoke with Nilüfer and her friend Banu in Istanbul on 1 July 2007.

35 Ilknur is 43 years old, a bank officer and lives in Ankara. I conducted an interview with her on 28 June 2007 in Ankara.

36 Yeşim is a university student who has recently moved to Ankara from Mersin for her education. She subsists on municipal and governmental student aid. I spoke with her on 23 June. 2007 in Ankara.

37 Kamile is a single mother who works as a tailor assistant at Nezaket's shop (whom I cite in chapter five). She is the only one who works in her family and struggles to feed her family. The interview took place in Ankara on 29 June 2007.

38 Kadir Şinas comments sarcastically on this kind of love: 'Have you ever lost your cell phone? I have and my world faded out…It is like death…My cellular was like a human being to me, it was witness of my life, my repository, my understanding friend…A new cellular means a new life, a new lover, and I am not ready for it' (*Radikal Iki*, 2000).

39 From the first interview that I conducted with Kaya which was on 22 June 2007.

40 Gaye lives in Istanbul and works as a manager at a private company. I spoke with her two times. The first one was on July 1 and the second one was on 15 July when there were some other participants discussing cellular telephony in Turkey.

41 Gülden (her real name) is a scriptwriter, working as a team member of a well-known creative group that produce popular television dramas in Turkey. She lives in Istanbul. Gülden and I also talked about the ways in which the cell phone has become one of the lead characters in television dramas in Turkey, particularly in detective stories. She told me that it is no longer possible to write an episode which does not include a scene where the characters use cell phones. She says, 'if your characters don't use cellular where they would use [it] if it was a real life [sic], then you lose the sense of reality. But to make our fiction more realistic, we produce dramas that are full of telephoning heads'. (2 July 2007/ Istanbul)

42 Here, I refer to Stephen David Ross' book on touch, in which he gathers and profoundly speaks of Heidegger, Levinas, and other philosophers' writings on the hand. See his *The Gift of Touch: Embodying The Good*, 1998.

43 I rely on Michel Serres' writings on touch, skin and hands. He points to Genesis: 'So what is a hand? It is not an organ, it is a faculty, a capacity for doing, for becoming claw or paw, weapon or compendium. It is a naked faculty. A faculty is not special, it is never specific, it is the possibility of doing something in general' (1995: 34-5).

44 The melancholic is also a fetishist in the sense that s/he always looks for an object that would substitute the loss with its presence and thereby sustain the phantasm (see chapter two). Agamben writes, 'covering its object with the funeral trappings of mourning, melancholy confers upon it the phantasmagorical reality of what is lost; but insofar as such mourning is for an unobtainable object, the strategy of melancholy opens a space for the existence of the unreal and marks out a scene in which the ego may enter into relation with it and attempt an appropriation such as no other possession could rival and no loss possibly threaten' (1993: 20).

5. INDIVIDUAL ARTICULATION WITH CELLULAR TELEPHONY: CONTAINMENT, TRANSFERENCE AND TRANSLATION

1 Although psychoanalysis as a discipline of thought delimits social analysis in its tendency to empty and fixate the social, historical, political and cultural contingencies into primordial understandings of 'guilt', 'lack', and the metaphysical understandings of subjectivity, it is still valuable for social analysis that aims to understand the relation between body and object (in a broader sense, including things, other bodies, the world). The body as we have mentioned earlier, by following Mauss, is a psychological, physical and social instrument and its actions as well as its relations and attachments need to be investigated with a triple approach which does not dismiss one of these aspects. Psychoanalysis is indeed promising for exploring the domain of the psyche, for interpreting the unconscious purposes, desires, wishes of the body that is attached to an object – which produces particular performances – and yet critical reading of psychoanalysis is necessary as well.

2 Sofia opposes the traditional view of considering spaces as passive constructs that do not participate in the articulation of bodies, things,

or objects. Her theory is based on approaching containing spaces as active in terms of articulating commonalities and differences between different entities in their containment or in gathering.

3 For instance, designating the femininity of containing technologies that have been neglected by the western philosophy of technology.

4 Sofia discusses the gathering performance of containing technologies in reference to Heidegger's philosophy of technology. While discussing the 'thingly' character of a jug, Heidegger draws attention to the containing dimension of the thing. He writes that the jug's character as a thing 'resides in its being qua vessel' (1971: 171) and Sofia adds, 'that is its capacity as a container' (191). After describing the jug's task as taking what is poured in and keeping what was poured, Heidegger suggests that this unity of 'two-fold containing' is 'determined by the outpouring for which the jug is fitted as jug' (ibid. 171-172 cited in Sofia, 191). For Heidegger the 'outpouring' is the gift, the donation that 'gathers what belong to giving' (ibid. 173). The outpoured gift is thereby interpreted as a gathering together of the various dimensions of containment.

5 Jonathan Sterne (2003a), describes articulation as 'the process by which different phenomena with no necessary relation to one another are made into a social unity' (183). He draws on Grossberg's description of articulation, in which Grossberg states: 'The concept of articulation provides a useful starting point for describing the process of forging connections between practices and effects, as well as of enabling practices to have different, often unpredicted effects. Articulation is the production of identity on top of difference, of unities out of fragments, of structures across practices' (1992: 54 cited in Sterne, 2003a: 183). Sterne continues, 'Each machine embodied a whole set of articulations; in turn, it was articulated to larger economic, technical, and social functions and relations among many other possible and actual uses' (183). Thus what gives character to the articulation of machines is largely the social context from which it emerges.

6 Psychoanalyst D. W. Winnicott (1971) spoke of the experience of relating to objects as the key element in cultural experience, and as critical to developing a sense of freedom.

7 The dialectical containing relation between mother and infant is originally the central idea of Wilfred Bion's theory. Ogden reviews Bion's theory along with Winnicott's theory of transitional phenomena in order to suggest that projective identification is a necessary process for an infant to become an individual. I will elaborate on these issues in

the forthcoming section. Yet at this point, it is significant to note that in Bion's, Winnicott's and also Ogden's discussions, individuation and becoming subject refer to the same process (i.e. individual articulation and becoming subject are identical in this context).

8 Ogden (1985) writes: 'specific forms of potential space include the play space, the area of the transitional object and phenomena, the analytical space, the area of cultural experience, and the area of creativity' (129).

9 Ronell (1989) writes that the mother's image sinks away in the use of the telephone; in reference to Heidegger's theory on Being, *Dasein*, and throwness she argues that it dissolves into an indeterminate They. She speaks about the indeterminate They in correlation with Lacan's (2002) big Other, which roughly describes the symbolic order – language – as the very force that initiates the subject.

10 Ogden suggests these in reference to Bion's notions of contained/container.

11 'The child had a woollen reel with a piece of string tied round it...what he did was to hold the reel by the string and very skillfully throw it over the edge of his curtained cot, so that it disappeared into it, at the same time uttering his expressive "o-o-o-o". He then pulled the reel out of the cot again by the string and hailed its reappearance with a joyful "*da*" [there]. This, then, was the complete game – disappearance and return. As a rule, one only witnessed its first act, which repeated untiringly as a game in itself though there is no doubt that the greater pleasure was attached to the second act' (Freud, 1990: 15).

12 Transference for Freud is essentially an act of repetition; this repetition is transference of a forgotten past, which for Freud is rooted in the subject's castration. As a field, psychoanalysis understands transference more generally as the reproduction of unconscious memories.

13 The mother and the self that are attached to the reel are not totally lost, even when it is out of sight, for the string is in the boy's hand. The ultimate pleasure is witnessing and mastering its re-appearance. Thus the whole play is dominated by Freud's pleasure principle, which he sees as the conditional principle of psychic life.

14 Ronell writes that the telephone becomes a contract for the boy's telephonic play of *fort/da*. She maintains this in reference to Derrida's interpretation of *fort/da* play where he writes: '[the child] makes himself re-, still in accordance with the law of the PP [pleasure principle, grandpa, i.e., Freud], in the grand speculation of a PP that seems never to leave himself/itself, nor anyone. This recalling by telephone or teletype (i.e., voice or writing, from afar) produces the "movement" by contracting

itself, by signing a contract with itself' (1978 cited in Ronell, 1995: 86).

15 Guattari (independently as well as in his works with Deleuze) has critiqued the psychoanalytic reduction of subjectivity into identifications of mommy-daddy images and formulations of Oedipus and castration – the imaginary and the symbolic – arguing that it 'systematically efface[s] the social, political and cultural contents of any psychic disturbance' (1995: 72). Directly confronting Lacanian psychoanalysis, he contends that 'psychoanalysis of the Lacanian stamp with its esoteric, pretentious character, cut off from all apprehension of the terrain of psychopathology entertains the idea that only individual treatment allows access to the "symbolic order" by transcendent routes of interpretation and transference' (ibid. 204). In a discussion of delirium, he notes that in all of his clinical observations, he has yet to encounter a case of delirium 'that is not firstly about race, racism, politics that does not begin in all directions from history, that does not involve culture' (ibid. 80-81). Elsewhere, in writing on machinism, he writes, 'machinism is an object of fascination, sometimes of delirium' (1990:13).

16 Turkcell's ad agency coined this word to refer to how cells appear as sweet and nice extra-humans. While the word 'cello' describes cellular technology, the word 'can' means the 'life' or 'heart' of living beings (including animals, plants and humans). 'Can' as a second name (such as Alican or Mehmetcan), is often given to babies, particularly those in the late 1990s and onward.

17 In one TV ad, a machinist on a long train trip misses his son and, because every Turkcell subscriber moves with a *cellocan* near to her/him, the machinist's insightful *cellocan* tells him to call his son. The machinist asks, 'How can I call my son, *cellocan*? We are in-between two mountains, it will not get the signal.' The *cellocan* tells him, 'It will, it will.' The machinist calls and is connected to his son, and says how much he misses his little boy. We then see the *cellocan* crying.

18 I talked with Savaş in Ankara, on 28. June. Savaş is a student at a public high school and he defines himself as 'a problem kid' who has no interest in school works and no dreams like going to a good university, having a decent job and wife for his future. His statements also appear in chapter six when I discuss cellular telephony as a network or fluid technology.

19 I reached Banu through Nilüfer and met her just after I conducted an interview with Nilüfer on 1 July in Istanbul. She did not give any personal information about her life. Her answers were short in general

and she did not seem very comfortable with speaking about the ways she uses her cellular in her life with a stranger. So I skipped most of my questions and kept the interview very short. Even her short statements have been very helpful in directing my thought from the issue of containment to translation.

20 Gerard Goggin writes, 'On his own blog, designer Christian Lindholm talks about the everyday things he does with Lifeblog and describes his cell phone as a "life recorder" (Lindholm, 2004 cited in Goggin, 2006: 160).

21 In Turkey, job opportunities for transvestites and transsexuals are severely limited. Even if one has sufficient qualifications for other occupations, there are not many opportunities. Thus transvestites and transsexuals are generally forced to work as sexual workers in order to feed themselves. One ways to find customers is to go out to the highways at night and wait for someone to stop, a practice that carries various risks. With a cell phone, as I learned from Nilufer, transvestite and transsexual workers are able to avoid such dangers, at least to a certain extent.

22 See Sterne, 2003a for the discussion of the institutionalization of the cell phone.

23 Sadie Plant (2000) identifies three categories for receiving a call in public. The first is flight, the user's immediate movement out of the group to talk in privacy. The second, suspension, is characterized by the user swiftly ending their action to take the call, often by not physically moving anywhere else. The third is persistence, which describes how the user tries to continue what s/he was doing prior to the call as much as possible and to stay in the place where s/he was. All three situations share a minimum disconnection from the user's immediate context (see also de Souza e Silva, 2006).

24 see www.hurriyet.com.tr/agore/article_main.asp?si=5.

25 According to TTK's 2006 research, the longest interaction (coupling) via cell phone lasts 15 minutes on average.

26 Similarly Haddon (2000) writes: 'the mobile clearly enables additional communication that we might have not before as does email – for example, phatic calls where the point is not so much the message, but the gesture of getting in touch.'

27 See Garcia-Montes et al. (2006), and Plant (2002) for further discussions of spontaneity and flexibility within the cell phone's performances.

28 I reached Nalan through Kaya who works at the same company with Kaya in Ankara. The above statements from her (Nalan) were taken

from the records of the interview on 22 June 2007.

29 Nokia announces, 'we are interested in other people's lives, but at the same time we want to share our own experiences and thoughts with the others. Having a multimedia diary of your life, is a unique starting point for sharing the items that make up your life memories' (cited in Goggin, 2006: 159).

30 Haraway writes: 'That's what articulation does, it is always a non-innocent, contestable practice, the partners are never set once and for all. There is no ventriloquism here. Articulation is work and it may fail. All the people, who care, cognitively, emotionally, and and politically must circulate their position in a field constrained by a new entity, made up of human and unhuman actors' (1992: 314).

31 I want to make clear that I do not use the term children as an implicit or explicit reference to the subjects of an immature Turkish modernity. In that respect, I do not consider cellular telephony as that which repairs the incomplete modernity of Turkey by giving the sense of modern subjects, moving and acting with speed. Turkish modernity, as Nilüfer Göle suggests about non-western modernities, does not 'lack' anything as a 'copy' of a western 'model'; its trajectory is different, and asserts its excess, its difference. In chapters two and three, I have outlined how I understand Turkish modern history in relation to melancholia and technoscape. The parallel I draw between 'third world' children and cellular telephony should be understood in reference to a melancholia that forecloses private worlds and social bonds.

6. CELLULUR TELEPHONY AS SOCIAL PRACTICE: THE COLLECTIVE DESIRE FOR LIVING IN AN OPEN CROWD

1 Ada used the Turkish word, *yığın* which can also be translated as 'mass'. However, his later explanation of *yığın* corresponds to the crowd as Bogard understands this term – where individual bodies feel that they are part of something while maintaining their singularity (see the last page of this chapter for Bogard's treatment of Canetti's conception of the crowd).

2 Ada is a 35 years old, male. He lives in Ankara and works as a freelance graphic designer. I conducted an interview with him on 28 June 2007.

3 *Alaturka* is a genre of traditional Turkish music.

4 For instance in the last national election, AKP (The Justice and Development Party) used cellular telephony as one of its primary

means of reaching many people with the voice message by party leader Recep Tayyip Erdogan (who became Prime Minister for a second time after the last national election in 2007). The communication manager of AKP's propaganda campaigns asserted on a television program that cellular telephony has been enormously effective in political campaigning. When one receives a message from the Prime Minister (he was the Prime Minister during the election campaigns) and hears his voice speaking directly to her/himself, one feels a proximity with a political leader whose public image is most often televisual.

5 A *dolmuş* is a sort of minibus, a common form of public transport in Turkey's cities and towns. Especially in Ankara, drivers take too many passengers, and some have to travel standing up. This is not easy, because the interior space of a *dolmuş* is not tall enough to stand in comfortably. For the *dolmuş* to work properly as an institution and as public transport, interaction between bodies is crucial. One needs to pass the money to the driver via other passengers, and to say loudly that one wants to get off in order to stop the car. The *dolmuş* is neither a bus whose interior design is organized in such a way that people can interact with buttons, ticket machines, or digital screens which display upcoming stops, nor is it a taxi whose route is determined by the customer and in which interaction occurs only between the driver and the passenger.

6 Civan comes from a family that is actively involved in Kurdish politics in Turkey. He lives both in rural and urban Ankara. I conducted an interview with him and his wife, Burcu, in Ankara, on 26 June 2007.

7 I am aware that Bruno Latour's concept of network differs in many ways from the network model I outline here. Yet, by following Mol and Law, I would like to highlight that even the shape of a network-object changes to a certain extent, as its use, performance, tasks, and meanings change. By emphasizing the possibility of change in an object's shape, I mean to stress how the fluidity of an object makes it open to appropriation in different localities, by responding to different collective desires, imaginations, and purposes. For this reason, thinking through shape is important in order to demonstrate that even the shape of an object can change alongside its domains of use, meanings, and performances.

8 The cell phone is also significant as a texting media for Kurds, whose language rights are severely limited in Turkey. Although there is always the risk of drawing state authorities' attention, one can send Kurdish SMS messages and circulate decentralized SMS messages through the net. The Kurdish alphabet is different from the Turkish (it contains

the letters w, and q, for instance) and language programs allow for a Kurdish subscriber to use her/his language for texting in daily life. Thus the cell phone is also ethnicized and collectivized in unexpected ways.

9 Turkey's previous Culture and Tourism minister became famous for his illness (as he calls it), which makes him sleepy when his body does not move. After being caught by television and newspaper cameras a number of times, he and his minister friends found a way to wake him up when he fell asleep in conferences or panels. Declaring that he would not be recorded sleeping in public meetings any more (see Zaman, 2007), the minister would put his cellular in vibrate mode, and as soon as he nodded off his guards would ring him. Thanks to the vibration of his phone, he would wake up and resume listening to the speaker. This has been a popular issue in Turkey for a while. (*Zaman*) [published on 07 October 2007].

10 Columnist Serdar Kuzuluoglu writes: 'Unfortunately, we have never been skilled in developing technologies. Moreover, we did not encourage those that want to do such things. So we have always tried to engage with technologies by looking at others who improve them. Fortunately, we have never disregarded technologies. Yet we have never lost the appetite to use it in our ways, in the way we do things here.' (*Radikal*, 2007) [published on 10 September 2007].

11 There is no actual speed-train system in Turkey. In 2004, the 'fast train' project – simply based on accelerating trains' velocity without changing the railways – caused a deadly crash; the existing tracks were unable to bear such high speeds. This crash, which took 34 lives, was reported with headlines such as 'An accelerated massacre' (*Sabah*), and 'They died in the name of show-off' (*Milliyet*) URL Document: [http://news.bbc.co.uk/1/hi/world/europe/3919787.stm] (accessed on 23 July 2004)

12 I spoke with the animator, Barış, who designed the Japanese man's flying cell phone in this ad. He told me that after this ad was broadcast on national TV channels, he came across some technological forum discussion boards where people debated whether such a telephone actually exists and if so when it could come to Turkey. He told me, 'I am sure if such a telephone was invented and introduced to the Turkish market, many would buy it right away…it's like we crave technological novelties…We were raised with American television series where we saw so many technological devices that we don't have in Turkey…I mean we have learned to desire these kinds of technologies…' (03 July 2007/ Istanbul).

13 I would like to note that I use the term 'globality' by following Sarah Ahmed (see chapter three) to emphasize the asymmetrical relations of nations, collectives, and individuals to the means of global flows and movement.

14 See also T. Moreira, 2004 in the discussion of circulation of such spaces in operation rooms.

15 For instance, while Telsim emphasizes the possibility that even 'third worlds' of this country can purchase, possess and adapt this technological novelty, Turkcell builds its brand-image on the perception that it is a network for Turkey's urbanized, modern segment, those who show an eagerness to progress towards modernity, and economic and technological development. While Telsim plays on nationalist tendencies (its logo is red and white, like the Turkish flag), Turkcell triggers the collective desire to be part of the global world (in Turkcell ads, Turkcell has been depicted as the one that not only catches up with the western counterparts but leave them behind).

16 For instance a Turkcell subscriber used to pay double the price when s/he coupled with a Telsim subscriber.

17 Tarık lives in Istanbul and works at a private company as a sales representative. I spoke with him on 3 July 2007 in Istanbul.

18 Avea is the cheapest one in the market now – Emine says that she often gives her Turkcell number to her customers in order to leave a good impression.

19 Specifically, a newspaper article on cellular telephony which states that telephone numbers that start with 2 are worth thousands of dollars: see Şükran Pakkan's article in *Milliyet Pazar*, 2004 [published on 22 February].

20 I asked a bank officer who is responsible for evaluating people's applications for bank loans, and she said to me that telephone numbers do not mean anything other than giving a contact number of the applicant, though she added later that if one does not have a cellular number, the bank often asks why (Ilknur, 25 June 2007/ Ankara).

21 Aycell and Aria entered the market in 2001, and they merged into the local-global brand of Avea in 2005.

22 Currently, when a Vodafone (Vodafone purchased Telsim in 2005) subscriber couples with Turkcell subscriber or Avea subscriber, s/he does not pay much more than s/he would to contact a Vodafone subscriber.

23 For instance, Vodafone – the old Telsim – has moved away from Telsim's marketing strategy which was based on an idea that can be

defined as 'enjoy your Turkishness', and in its new approach 'seizing the moment', especially the moment of global time, is celebrated.

24 Süreyya is a middle-aged, married woman who used to be involved in the left-wing student movement in the 1970s and 1980s. Just as other 'revolutionaries' (as they call themselves) of the time, for her 'the self' is only meaningful when it is devoted to a social and political project which would change the existing conditions.

25 Burcak is a high school student and lives in Ankara. He is the son of Nezaket whom I have cited before in chapter five. I spoke with him on 26 June in Ankara.

26 I spoke with Papatya in Istanbul on 3 August 2007. Papatya is a television celebrity who produces television shows in Turkey. I wanted to conduct an interview with her to ask how a celebrity uses a cell phone, but our talk turned out to flow in a somewhat different direction.

27 Jean Baudrillard immediately springs to mind, as he steadfastly maintained that 'contacts' through media have replaced the sensuousness of touch. He believed that we are tested everywhere, and that the method for this test is tactical. In the realm of communication, where the sensory value of touch is lost to simulation, all forms of social rapport are transformed into 'contacts' and an entire social configuration orbits around the test – the question and answer call (1983). With the presupposition that before haptic technologies, there were forms of social rapport and social bonds in social spheres, one can argue that cellular telephonic touch, which is itself a haptic technology that is meant to produce a sensory experience of social relations, takes the place of sensory touch between bodies.

28 Meyda is a veiled university student who is involved in an Islamist student movement. Her family and most of her friends live in Sivas. I conducted an interview with her in Ankara on 23 June 2007.

29 Nihat is a middle-aged man. I reached him through Nezaket and conducted an interview with him on 27 June in Ankara.

INDEX